TECHNICAL REPORT

U.S. Combat Commands' Participation in the Proliferation Security Initiative

A Training Manual

Charles Wolf, Jr., Brian G. Chow, Gregory S. Jones

Prepared for the Office of the Secretary of Defense

NATIONAL DEFENSE RESEARCH INSTITUTE

The research described in this report was prepared for the Office of the Secretary of Defense (OSD). The research was conducted in the RAND National Defense Research Institute, a federally funded research and development center sponsored by the OSD, the Joint Staff, the Unified Combatant Commands, the Department of the Navy, the Marine Corps, the defense agencies, and the defense Intelligence Community under Contract W74V8H-06-C-0002.

Library of Congress Cataloging-in-Publication Data is available for this publication.

ISBN 978-0-8330-4696-3

The RAND Corporation is a nonprofit research organization providing objective analysis and effective solutions that address the challenges facing the public and private sectors around the world. RAND's publications do not necessarily reflect the opinions of its research clients and sponsors.

RAND® is a registered trademark.

Published 2009 by the RAND Corporation
1776 Main Street, P.O. Box 2138, Santa Monica, CA 90407-2138
1200 South Hayes Street, Arlington, VA 22202-5050
4570 Fifth Avenue, Suite 600, Pittsburgh, PA 15213-2665
RAND URL: http://www.rand.org/
To order RAND documents or to obtain additional information, contact
Distribution Services: Telephone: (310) 451-7002;
Fax: (310) 451-6915; Email: order@rand.org

Preface

As one of two parts of the RAND Corporation's recent work on the Proliferation Security Initiative (PSI), RAND developed a manual for the Geographic Combat Commands (GCCs) to use in training personnel assigned to GCCs for participation in PSI exercises. It was felt that these training materials would help the GCCs deal with normal issues arising from staff turnover and sometimes insufficient institutional memory. Since PSI's inception, in 2003, there have been 36 of these exercises, which constitute the core of PSI's regular, sustained activities.

This training manual is configured as nine sessions of lectures and seminars. The material in this manual draws from and contributes to the document describing the other part of RAND's recent work on PSI: *Enhancement by Enlargement: The Proliferation Security Initiative*, MG-806-OSD, 2008, by Charles Wolf, Jr., Brian G. Chow, and Gregory S. Jones.

This research was sponsored by the Office of the Secretary of Defense and conducted within the International Security and Defense Policy Center of the RAND Corporation's National Defense Research Institute, a federally funded research and development center sponsored by the Office of the Secretary of Defense, the Joint Staff, the Unified Combatant Commands, the Department of the Navy, the Marine Corps, the defense agencies, and the defense intelligence community.

For more information on RAND's International Security and Defense Policy Center, contact the Director, James Dobbins. He can be reached by email at James_Dobbins@rand.org; by phone at 703-413-1100, extension 5134; or by mail at the RAND Corporation, 1200 S. Hayes Street, Arlington, VA 22202. More information about RAND is available at www.rand.org.

Contents

Summary

This document is a manual for the Geographic Combat Commands (GCCs) to use in training personnel assigned to GCCs for participation in the Proliferation Security Initiative (PSI) exercises. Its purpose is to help the GCCs deal with the normal issues arising from staff turnover and sometimes insufficient institutional memory. Since the inception of the Proliferation Security Initiative (PSI) in 2003,[1] 36 of these exercises, which constitute the core of PSI's regular, sustained activities, have been planned and carried out.

This training manual[2] is one of two documents covering RAND's recent work on PSI for the Office of the Secretary of Defense (OSD). The companion document, which both draws from and contributes to the material in this manual, focuses on the enhancement of PSI's effectiveness through its enlargement to include five key countries that have so far chosen not to participate in PSI (i.e., the "holdout" countries).[3]

This manual consists of nine sessions of lectures and seminars, each programmed for one or two hours. The sessions are structured so that the number of sessions used and/or the time allocated to an individual session can be reduced to fit the GCC's training purpose and available time. The material covered in the nine sessions is as follows:

- *Session 1: PSI History and Background.* This session addresses PSI's creation in 2003, its purpose, its role as an *activity* (not an organization or an international agreement) aimed at preventing the spread of weapons of mass destruction (WMD), and its conduct of interdiction training exercises.
- *Session 2: PSI Design and Interdiction Principles.* The focus in this session is on how PSI works to serve its purpose, the central role of PSI's Operational Experts Group (OEG), and the basic interdiction principles that provide the basis for voluntary cooperation among the 93 countries affiliated with PSI.
- *Session 3: U.S. Laws Relevant to PSI.* This session concentrates on the legal basis for the military support that the United States provides to PSI, as well as on the criminal and civil legal infrastructure authorizing intelligence collection, export control, and border control among the activities embraced by PSI.

---

[1]   See U.S. Department of State, Under Secretary for Arms Control and International Security, Bureau of International Security and Nonproliferation, *Proliferation Security Initiative (PSI)*, Fact Sheet, Bureau of International Security and Nonproliferation, Washington, D.C., May 26, 2008.

[2]   Originally planned as a syllabus, this document evolved into a training manual over the course of the project.

[3]   The companion document is Charles Wolf, Jr., Brian G. Chow, and Gregory S. Jones, *Enhancement by Enlargement: Proliferation Security Initiative*, MG-806-OSD, Santa Monica, Calif.: RAND, 2008.

- *Session 4: International Agreements Relevant to PSI.* This session addresses the various international agreements that provide legitimacy for PSI by making nonproliferation a universal norm. The agreements are briefly described in groups covering counterterrorism, United Nations Security Council Resolutions, bilateral ship-boarding agreements, the United Nations Convention on the Law of the Sea, and U.S.-sponsored programs that assist other countries in detecting concealed WMD items.

- *Session 5: Incentives and Disincentives for PSI Participation.* This session focuses on the incentives and disincentives that have figured in the choice made by more than 90 countries to affiliate with PSI and in the choice of the five "holdout" countries (China, India, Pakistan, Indonesia, and Malaysia) to refrain from affiliation. Both the public-good and the collective benefits resulting from PSI are discussed, along with the national benefits accruing to PSI affiliates. Also considered are the disincentives that some countries associate with PSI: possible abridgement of their sovereignty, compromise of their independent foreign policy, limits on the right of innocent passage, and possible violation of the law of the sea.

- *Session 6: Detection of WMD, Their Delivery Systems, and Related Materials.* This session covers the detection of illicit *WMD items*, by which we mean WMD, their delivery systems, and related materials. The WMD items are grouped according to whether they are radioactive or nonradioactive (this second category includes chemical and biological weapons and agents), and the different systems and techniques for detecting WMD items are then discussed separately for the two groups.

- *Session 7: PSI Exercises and Lessons Learned.* This session addresses the what, when, and who questions associated with the 36 exercises that PSI activities have encompassed since 2003—the effect that these exercises have had on the will and determination of nations to counter proliferation, the strength of and coordination among the countries and agencies participating in the exercises, and the expanded range of countries that engage in the exercises.

- *Session 8: Responding to Issues Challenging PSI.* This session considers several challenges and objections to PSI that have arisen. The challenges and objections and appropriate responses to them are discussed in relation to specific topics: the law of the sea, the right of innocent passage, uncertainty about the circumstances in which PSI interdiction efforts would actually be applied, and the putative U.S. dominance of PSI that causes some countries to be concerned that PSI affiliation will imply closer association with U.S. policies than they would like.

- *Session 9: Enhancing Capabilities for PSI Participation.* A nation's affiliation with PSI— and the frequency and intensity of its participation—is entirely voluntary. This session is concerned with the ways in which PSI's exercises and discussions can affect the capabilities of participating countries; in other words, how participation can improve customs and invoicing practices, enhance inspection and detection capabilities, increase the sharing of information related to suspected proliferation activities, increase the interoperability of communications and other systems, improve interdiction and decisionmaking processes, and aid in identifying and, where necessary, interdicting transshipment of WMD items.

Note to the Reader

Between the writing of this volume and that of the earlier, companion volume, changes occurred that affect some information relevant to both volumes: (1) There are now 93 countries, rather than 91, participating in PSI. (2) The number of PSI exercises that have been conducted is now 36, not 34. (3) Documents formally located on U.S. government Web sites have ceased to be at those sites, primarily because they have been moved to archival locations in reflection of the change in the U.S. administration that occurred on January 20, 2009. The numbers, documents, and URLs in this training manual are current as of February 2009.

Acknowledgments

We are pleased to acknowledge the useful comments we received from RAND colleague David Mosher and Stanford's Professor Harry Rowen on an earlier version of this report.

Abbreviations

AECA	Arms Export Control Act
AG	Australia Group
AOR	area of responsibility
BTWC (also BWC)	Convention on the Prohibition of the Development, Production and Stockpiling of Bacteriological (Biological) and Toxin Weapons and on Their Destruction
CBM	confidence building measure
CBP	Customs and Border Protection
CI	counterintelligence
CIT	Commodity Identification Training
CJCS	Chairman of the Joint Chiefs of Staff
CSI	Container Security Initiative
CWC	Convention on the Prohibition of the Development, Production, Stockpiling and Use of Chemical Weapons and on Their Destruction
CZT	cadmium-zinc-tellurium
DHS	Department of Homeland Security
DoD	Department of Defense
EAA	Export Administration Act
EAR	Export Administration Regulations
EU	European Union
EXBS	Export Control and Related Border Security Assistance
FAQ	frequently asked questions
FBI	Federal Bureau of Investigation
GCC	Geographic Combat Command
HEU	highly enriched uranium
IAEA	International Atomic Energy Agency
ICE	Immigration and Customs Enforcement

MTCR	Missile Technology Control Regime
NaI	sodium iodide
NATO	North Atlantic Treaty Organization
NSPD	National Security Presidential Directive
OEG	Operational Experts Group
OSD	Office of the Secretary of Defense
PFNA	Pulsed fast neutron analysis
PSI	Proliferation Security Initiative
Pu	Plutonium
SCO	Shanghai Cooperation Organization
SFI	Security Freight Initiative
SME	subject-matter expert
SNM	special nuclear material
TSA	Transportation Security Administration
U	Uranium
UN	United Nations
UNSCR	United Nations Security Council Resolution
USCG	U.S. Coast Guard
USJFCOM	U.S. Joint Forces Command
USSTRATCOM	U.S. Strategic Command
WMD	weapons of mass destruction

PSI History and Background

PSI History

President George W. Bush announced the creation of the Proliferation Security Initiative (PSI) in Krakow, Poland, on May 31, 2003. The announcement was brief:

> When weapons of mass destruction or their components are in transit, we must have the means and authority to seize them. So today I announce a new effort to fight proliferation called the Proliferation Security Initiative. The United States and a number of our close allies, including Poland, have begun working on new agreements to search planes and ships carrying suspect cargo and to seize illegal weapons or missile technologies. Over time, we will extend this partnership as broadly as possible to keep the world's most destructive weapons away from our shores and out of the hands of common enemies.[1]

Meetings were held in Madrid, Spain, on June 12, 2003, and in Brisbane, Australia, on July 9 and 10, 2003, leading to a meeting in Paris, France, on September 3 and 4, 2003. A key outcome of this last meeting was the adoption of a "statement of interdiction principles."[2] Initially, 11 countries endorsed these principles: Australia, France, Germany, Italy, Japan, the Netherlands, Poland, Portugal, Spain, the United Kingdom, and the United States.

The interdiction principles are summarized in the following paragraphs (see Session 2, PSI Design and Interdiction Principles, for a detailed discussion).[3]

The main purpose of the PSI interdiction principles is to "establish a more coordinated and effective basis through which to impede and stop shipments of WMD [weapons of mass destruction], delivery systems, and related materials flowing to and from states and non-state actors of proliferation concern." Actions taken to carry out this objective are to be "consistent with national legal authorities and relevant international law and frameworks, including the UN Security Council."

As part of the interdiction principles, countries are to "adopt streamlined procedures for rapid exchange of relevant information," "review and work to strengthen their relevant national legal authorities where necessary," "work to strengthen when necessary relevant international

---

[1]   George W. Bush, speech given in Krakow, Poland, May 31, 2003. Full transcript available in "Bush Urges NATO Nations to Unite in Fight Against Terrorism," May 31, 2003.

[2]   *Proliferation Security Initiative: Statement of Interdiction Principles*, Paris, September 4, 2003.

[3]   All quotations in this discussion, unless otherwise noted, are from U.S. Department of State, Under Secretary for Arms Control and International Security, Bureau of International Security and Nonproliferation, *Interdiction Principles for the Proliferation Security Initiative*, Bureau of International Security and Nonproliferation, Washington, D.C., September 4, 2003.

law and frameworks," and "take specific actions in support of interdiction efforts." In this last category, countries are "not to transport or assist in the transport of any such cargoes"; "to take action to board and search any vessel flying their flag"; "to seriously consider providing consent under the appropriate circumstances to the boarding and searching of [their] own flag vessels by other states"; "to stop and/or search in their internal waters, territorial seas, or contiguous zones . . . vessels that are reasonably suspected of carrying such cargoes"; "to . . . require aircraft . . . that are transiting their airspace to land for inspection"; and "if their ports, airfields, or other facilities are used as transshipment points . . . , to inspect vessels, aircraft, or other modes of transport."

PSI is not an international agreement. It is "an innovative and proactive approach to preventing proliferation that relies on voluntary actions by states," and it

> provides a basis for cooperation among partners on specific actions when the need arises. Interdictions are information-driven and may involve one or several participating states, as geography and circumstances require. The PSI is not a formal treaty-based organization, so it does not obligate participating states to take specific actions at certain times. By working together, PSI partners combine their capabilities to deter and stop proliferation wherever and whenever it takes place."[4]

In addition to conducting interdictions, PSI members participate in training exercises: "A robust PSI exercise program allows participants [to] increase their interoperability, improve interdiction decision-making processes, and enhance the interdiction capacities and readiness of all participating states."[5] As of January 22, 2009, 93 countries had endorsed the PSI interdiction principles.[6]

To facilitate interdictions under PSI, the United States has signed ship-boarding agreements with nine countries: Bahamas, Belize, Croatia, Cyprus, Liberia, Malta, Marshall Islands, Mongolia, and Panama.[7] These agreements are modeled after similar arrangements that exist in the counternarcotics arena. They provide authority on a bilateral basis to board ships registered under the flag of one of these nine countries and believed to be carrying suspect cargoes. They establish procedures to board and search vessels in international waters. Under these agreements a vessel may be searched after as little as two hours after a request has been made by a third party. These nine countries are ones that shippers often use as "flags of convenience" and have the majority of the world's shipping operating under their flags.

---

[4]   U.S. Department of State, Under Secretary for Arms Control and International Security, Bureau of International Security and Nonproliferation, *Proliferation Security Initiative (PSI)*, Fact Sheet, Bureau of International Security and Nonproliferation, Washington, D.C., May 26, 2008.

[5]   U.S. Department of State, Under Secretary for Arms Control and International Security, Bureau of International Security and Nonproliferation, *Proliferation Security Initiative (PSI)*, 2008.

[6]   U.S. Department of State, Under Secretary for Arms Control and International Security, Bureau of International Security and Nonproliferation, "Proliferation Security Initiative Participants," Web page, current as of January 22, 2009.

[7]   U.S. Department of State, Under Secretary for Arms Control and International Security, Bureau of International Security and Nonproliferation, "Ship Boarding Agreements," Web page with description and links to U.S.-country agreements (e.g., "Proliferation Security Initiative Ship Boarding Agreement with Belize"), undated.

In the early years of PSI, a "core" group of member countries defined the basic principles of interdiction and worked to expand support.[8] This group was disbanded in August 2005 after India (which has not endorsed the PSI interdiction principles) complained of discrimination among PSI participants. Instead, there is now the PSI Operational Experts Group (OEG), which is

> a group of military, law enforcement, intelligence, legal, and diplomatic experts from twenty PSI participating states [that] meets regularly to develop operational concepts, organize the interdiction exercise program, share information about national legal authorities, and pursue cooperation with key industry sectors. The OEG works on behalf of *all* PSI partners and works enthusiastically to share its insights and experiences through bilateral and multilateral outreach efforts.[9]

The 20 members of this group are Argentina, Australia, Canada, Denmark, France, Germany, Greece, Italy, Japan, the Netherlands, New Zealand, Norway, Poland, Portugal, Russia, Singapore, Spain, Turkey, the United Kingdom, and the United States.

UNSCR 1540

United Nations Security Council Resolution (UNSCR) 1540,[10] which was adopted on April 28, 2004, has a purpose similar to that of PSI. It calls on all states to "refrain from providing any form of support to non-State actors that attempt to develop, acquire, manufacture, possess, transport, transfer or use nuclear, chemical or biological weapons and their means of delivery". Note that the PSI interdiction principles refer to both state and non-state actors, whereas UNSCR 1540 refers only to non-state actors.

UNSCR 1540 calls on all states to adopt and enforce laws and various measures to achieve this goal. It also calls on all states to submit a report to the UN Security Council on the implementation of UNSCR 1540. As of December 2004, reports had been received from 87 states and the European Union (EU).

It has been reported that the original purpose of UNSCR 1540 was to endorse PSI and to provide authority for interdiction of ships on the high seas. However, it has also been reported that because of a threatened Chinese veto, the current version makes no mention of PSI and provides no authority for the interdiction of ships on the high seas.[11]

---

[8]　Sharon Squassoni, "Proliferation Security Initiative (PSI)," *CRS Report for Congress*, RS21881, September 14, 2006, p. 2.

[9]　U.S. Department of State, Under Secretary for Arms Control and International Security, Bureau of International Security and Nonproliferation, *Proliferation Security Initiative (PSI)*, 2008.

[10]　UNSCR 1540 (2004), S/RES/1540 (2004), April 28, 2004.

[11]　William Hawkins, "Chinese Realpolitik and the Proliferation Security Initiative," February 18, 2005.

PSI Background and Ship Interdictions

Part of the groundwork for PSI was established in December 2002 with the White House release of National Security Presidential Directive (NSPD) 17, *National Strategy to Combat Weapons of Mass Destruction*.[12] NSPD 17 describes three "pillars of our national strategy," one of which is "counterproliferation to combat WMD use."; it then lists "interdiction" as one of three capabilities needed for counterproliferation stating that "[e]ffective interdiction is a critical part of the U.S. strategy to combat WMD and their delivery means."

An incident that occurred around the same time as the release of this report provided further impetus for the creation of PSI. A ship, the *So San*, sailed from North Korea into the Gulf of Aden. The United States had been aware of the ship almost since its departure from North Korea. There was concern that the ship was carrying "weapons of concern," and the U.S. Navy had monitored the vessel throughout its journey.[13] On December 9, 2002, the ship was intercepted by the Spanish frigate *Navarra* at the request of the United States. The ship was not flying a flag, refused a request for boarding, and accelerated. The ship was boarded by Spanish marines using a helicopter. The ship was registered in Cambodia and had a North Korean crew. The limited paperwork on the ship indicated that it was carrying a cargo of cement to Djibouti. However, a search of the ship found 15 complete Scud missiles and 23 containers of nitric acid hidden under the sacks of cement. The nitric acid is used as part of the missile's propellant. When a U.S. ship started to escort the *So San* to Diego Garcia, Yemeni officials protested to the U.S and Spanish governments. The consignment of missiles was bound for Yemen. Yemeni officials could not explain why the missiles were hidden and why the ship did not have the proper paperwork. After a few days, the United States allowed the ship to deliver its cargo to Yemen, though the decision puzzled Spanish authorities. It has been speculated that U.S. desire for Yemen's assistance in the war on terror led to the U.S. decision. One result of this affair was that the White House spokesman, Ari Fleischer, indicated that this incident showed the need for additional international anti-proliferation measures.

Another ship interdiction illustrates the potential benefits of PSI.[14] In October 2003, the *BBC China* was sailing from Dubai to Tripoli, Libya. It was flying a flag of convenience (Antigua and Barbuda) and was owned by a German shipping company. At the request of British and U.S. authorities, German authorities asked the shipping company to voluntarily divert the ship to Italy, where it was searched. On board were a large number of centrifuge components intended for a clandestine uranium enrichment plant in Libya. This equipment was seized before the vessel was allowed to complete its voyage. On December 19, 2003, Libya announced that it would dismantle its WMD programs, disclose all relevant information about those programs, and allow inspectors to verify its compliance.[15] As part of this process, the full extent of the illicit arms network run by the Pakistani A.Q. Khan was revealed. It is not clear how much of a role this ship interdiction played in Libya's decision to give up its WMD program. Apparently, Libya was already considering this action before the *BBC China* seizure, but this

---

[12]   *National Strategy to Combat Weapons of Mass Destruction*, NSPD 17, December 2002.

[13]   Brian Knowlton, "Ship Allowed to Take North Korea Scuds on to Yemeni Port: U.S. Frees Freighter Carrying Missiles," *International Herald Tribune*, December 12, 2002.

[14]   Stockholm International Peace Research Institute, *SIPRI Yearbook 2005: Armaments, Disarmament and International Security*, United Kingdom: Oxford University Press, 2005, pp. 640 and 748.

[15]   Paul Kerr, "Libya Vows to Dismantle WMD Program," *Arms Control Today*, January/February 2004, p. 29.

event certainly must have helped Libya finalize its decision. This event was initially touted as one of PSI's successes, but government officials later indicated that the investigation into the Khan network was already ongoing when PSI was created.[16] Nevertheless, the interdiction of the *BBC China* illustrates the benefits that can be achieved through PSI actions.

The interdiction of the Chinese ship *Yin He*, which occurred nearly a decade before PSI's creation, has unfavorably colored China's view of PSI.[17] On July 15, 1993, the *Yin He* left China for various ports in the Middle East. U.S. intelligence had information that the ship was carrying the chemicals thiodiglycol and thionyl chloride, which can be used to manufacture mustard gas, a chemical warfare agent. It was believed that these chemicals would be delivered to the Iranian port of Bandar Abbas. The United States delivered a démarche to China on this matter on July 23, and U.S. naval vessels began to shadow the vessel. High-level Chinese officials denied that the chemicals were on board and offered to have the ship inspected in a neutral port. The United States believed that China was bluffing and made it clear that the ship would not be allowed to dock at Bandar Abbas. Other countries in the Persian Gulf region would not allow the *Yin He* to dock in their ports for inspection, and the ship spent several weeks anchored at various points in the Persian Gulf while the rhetoric between the United States and China escalated. Finally, Saudi Arabia allowed the ship to dock for inspection at Dammam. The inspection, which took place from August 26 to September 4, 1993, was nominally carried out by Saudi Arabia, but a number of Americans took part in the inspection to help guide the Saudis. The inspection report stated: "The complete inspection of all the containers aboard the 'YIN HE' showed conclusively that the two chemicals, thiodiglycol and thionyl chloride, were not among the ship's cargo."[18] China continues to be resentful of this incident and it is likely to view it as an impediment to participation in PSI.

This incident shows that caution is needed in using intelligence information for PSI purposes. However, it needs to be remembered that setting too high of a standard risks failing to interdict dangerous transfers.

In 1999, another interdiction of interest occurred.[19] On June 25 of that year, the North Korean freighter *Kuwolsan* docked at the Indian port of Kandla to unload a consignment of sugar. Acting on a tip, India customs agents attempted to board and search the ship. The North Korean crew physically impeded them but eventually, at the threat of gunpoint, relented. Because the customs officials lacked the expertise needed, experts from Indian's missile establishment were called in to examine the items found.[20] Hidden inside crates labeled "water refinement equipment" was what has been termed "an entire assembly line for missiles." This included machine tools, guidance systems, and engineering drawings labeled "Scud B" and "Scud C." The Indians seized all of these items despite North Korean protests. Ironically, the *Kuwolsan* was not supposed to be traveling to India at all. The ship's captain, in an attempt to earn extra money, had picked up a load of sugar in Thailand to sell in Algeria, on the way to

---

[16] Wade Boese, "Key U.S. Interdiction Initiative Claim Misrepresented," *Arms Control Today*, July/August 2005, pp. 26–27.

[17] Robert L. Suettinger, *Beyond Tiananmen: The Politics of U.S.-China Relations, 1989–2000*, Brookings Institution Press, Washington D.C., 2003, pp. 174–177.

[18] Ministry of Foreign Affairs of the People's Republic of China, "Statement by the Ministry of Foreign Affairs of the People's Republic of China on the 'Yin He' Incident, dated 4 September 1993."

[19] Joby Warrick, "On North Korean Freighter, a Hidden Missile Factory," *Washington Post*, August 14, 2003.

[20] "Customs Seek Help of Experts," *Indian Express*, July 1, 1999.

deliver the primary shipment. When the Algerian deal fell through after the sugar had been picked up, he arranged to deliver it to India instead. India believed that Pakistan was the intended recipient of the shipment, but Libya is also a possibility. This incident illustrates the extent of the traded items that PSI is intended to control and shows the power that states have over ships in their ports because internal waters are sovereign territory. It also illustrates that ordinary customs officials do not have the expertise needed to evaluate items of concern to PSI and thus will need the help of experts in these fields.

Readings for Session 1: PSI History and Background

Mayuka Yamazaki, *Origin, Developments and Prospects for the Proliferation Security Initiative*, Institute for the Study of Diplomacy, Edmund A. Walsh School of Foreign Service, Georgetown University, Washington, D.C., 2006. Downloadable as of February 17, 2009: http://isd.georgetown.edu/JFD_2006_PSA_Yamazaki.pdf
This article provides a good background on the origins and early years of PSI.

Andrew C. Winner, "The Proliferation Security Initiative: The New Face of Interdiction," *Washington Quarterly*, Spring 2005. Downloads of February 19, 2009: http://www.twq.com/05spring/docs/05spring_winner.pdf
This article provides a view of PSI from someone who is a strong supporter.

Mark J. Valencia, "The Proliferation Security Initiative: A Glass Half-Full," *Arms Control Today*, June 2007. As of February 19, 2009: http://www.armscontrol.org/act/2007_06/Valencia
This article takes a broad historical view of ship interdiction and how the law in this area has changed over time.

Samuel E. Logan, "The Proliferation Security Initiative: Navigating the Legal Challenges," *Journal of Transnational Law & Policy*, 14(2), Spring 2005. Downloadable as of February 28, 2009, at: http://www.law.fsu.edu/journals/transnational/backissues/issue14_2.html
This article examines the legal issues limiting PSI and discusses possible legal approaches to overcoming these limitations.

PSI Design and Interdiction Principles

PSI Design

The U.S. Department of State's fact sheet on the Proliferation Security Initiative (PSI) provides the basic information on PSI's design:[1] "The Proliferation Security Initiative (PSI) is a global effort that aims to stop trafficking of weapons of mass destruction (WMD), their delivery systems, and related materials to and from states and non-state actors of proliferation concern." Additionally: "The PSI is an innovative and proactive approach to preventing proliferation that relies on voluntary actions by states that are consistent with national legal authorities and relevant international law and frameworks. PSI participants use existing authorities—national and international—to put an end to WMD-related trafficking and take steps to strengthen those authorities as necessary."

Particularly useful is the section describing how PSI works:

> The PSI works in three primary ways. First, it channels international commitment to stopping WMD-related proliferation by focusing on interdiction as a key component of a global counterproliferation strategy. Endorsing the PSI Statement of Interdiction Principles provides a common view of the proliferation problem and a shared vision for addressing it.

> Second, the PSI provides participating countries with opportunities to improve national capabilities and authorities to conduct interdictions. A robust PSI exercise program allows participants increase their interoperability, improve interdiction decision-making processes, and enhance the interdiction capacities and readiness of all participating states. In five years, PSI partners have sustained one of the only global, interagency, and multinational exercise programs, conducting over 30 operational air, maritime, and ground interdiction exercises involving over 70 nations. These exercises are hosted throughout the world by individual PSI participants and consist of air, maritime, and ground exercises executed by participants' interagency and ministries focusing on improving coordination mechanisms to support interdiction-related decision-making.

> Furthermore, the PSI Operational Experts Group (OEG), a group of military, law enforcement, intelligence, legal, and diplomatic experts from twenty PSI participating states, meets regularly to develop operational concepts, organize the interdiction exercise program, share information about national legal authorities, and pursue cooperation with key industry sec-

---

[1]   U.S. Department of State, Under Secretary for Arms Control and International Security, Bureau of International Security and Nonproliferation, *Proliferation Security Initiative (PSI)*, Fact Sheet, Bureau of International Security and Nonproliferation, Washington, D.C., May 26, 2008. Unless otherwise noted, quotations in the discussion are from this source.

tors. The OEG works on behalf of *all* PSI partners and works enthusiastically to share its insights and experiences through bilateral and multilateral outreach efforts.

Third, and of the most immediate importance, the PSI provides a basis for cooperation among partners on specific actions when the need arises. Interdictions are information-driven and may involve one or several participating states, as geography and circumstances require. The PSI is not a formal treaty-based organization, so it does not obligate participating states to take specific actions at certain times. By working together, PSI partners combine their capabilities to deter and stop proliferation wherever and whenever it takes place.

Endorsement of the PSI interdiction principles is a key step for any country wishing to participate in PSI. These principles are discussed in detail below.

The OEG consists of 20 countries: Argentina, Australia, Canada, Denmark, France, Germany, Greece, Italy, Japan, Netherlands, New Zealand, Norway, Poland, Portugal, Russia, Singapore, Spain, Turkey, United Kingdom, and United States.

Note that the fact sheet calls PSI not "a formal treaty-based organization" but an "activity." Also note that endorsement of the PSI interdiction principles does not obligate participating countries to take any specific action and that a country's decision to participate in an interdiction is completely voluntary.

PSI Interdiction Principles

This section presents the introduction to the PSI interdiction principles and the principles themselves, along with annotation.[2]

Interdiction Principles for the Proliferation Security Initiative

PSI participants are committed to the following interdiction principles to establish a more coordinated and effective basis through which to impede and stop shipments of WMD, delivery systems, and related materials flowing to and from states and non-state actors of proliferation concern, consistent with national legal authorities and relevant international law and frameworks, including the UN Security Council. They call on all states concerned with this threat to international peace and security to join in similarly committing to:

1. Undertake effective measures, either alone or in concert with other states, for interdicting the transfer or transport of WMD, their delivery systems, and related materials to and from states and non-state actors of proliferation concern. "States or non-state actors of proliferation concern" generally refers to those countries or entities that the PSI participants involved establish should be subject to interdiction activities because they are engaged in proliferation through: (1) efforts to develop or acquire chemical, biological, or nuclear weapons and associated delivery systems; or (2) transfers (either selling, receiving, or facilitating) of WMD, their delivery systems, or related materials.

2   All quotations in this discussion are from U.S. Department of State, Under Secretary for Arms Control and International Security, Bureau of International Security and Nonproliferation, *Interdiction Principles for the Proliferation Security Initiative*, Bureau of International Security and Nonproliferation, Washington, D.C., September 4, 2003.

This introduction and first provision lay out the basic purpose of PSI—the interdiction of WMD (chemical, biological, or nuclear weapons), their delivery systems, and related materials. They indicate that PSI is to work within national legal authorities and relevant international law and that PSI is directed at "states or non-state actors of proliferation concern." Exactly which countries are meant by this phrase is not indicated, but it is stated that this determination will be made by the PSI participants. However, since PSI has no organizational structure, it is not clear how this will be done. Presumably, such countries as North Korea, Iran, and Syria are likely fits for the "of proliferation concern" category, but other countries, such as Pakistan and India, have been concerned that PSI might be directed against them as well, which is one reason behind their reluctance to participate in PSI.

> 2. Adopt streamlined procedures for rapid exchange of relevant information concerning suspected proliferation activity, protecting the confidential character of classified information provided by other states as part of this initiative, dedicate appropriate resources and efforts to interdiction operations and capabilities, and maximize coordination among participants in interdiction efforts.

This second provision is probably the least controversial of the lot. The extent to which information exchange takes place is hard to calibrate because exchanges can occur without any actions that would be visible to those outside of government.

> 3. Review and work to strengthen their relevant national legal authorities where necessary to accomplish these objectives, and work to strengthen when necessary relevant international law and frameworks in appropriate ways to support these commitments.

Provision 3 received reinforcement from the adoption on April 28, 2004, of UN Security Council Resolution (UNSCR) 1540,[3] which called on all states to "refrain from providing any form of support to non-State actors that attempt to develop, acquire, manufacture, possess, transport, transfer or use nuclear, chemical or biological weapons and their means of delivery." UNSCR 1540 calls on all states to adopt and enforce laws and various measures to achieve this goal and calls on all states to submit a report to the Security Council on the resolution's implementation. Note that the PSI interdiction principles refer to both state and non-state actors, whereas UNSCR 1540 refers only to non-state actors.

> 4. Take specific actions in support of interdiction efforts regarding cargoes of WMD, their delivery systems, or related materials, to the extent their national legal authorities permit and consistent with their obligations under international law and frameworks, to include:
>
> a. Not to transport or assist in the transport of any such cargoes to or from states or non-state actors of proliferation concern, and not to allow any persons subject to their jurisdiction to do so.

Provision 4 spells out actions to be taken for the interdiction of WMD cargoes. The introductory portion and Section (a) restate the basic purpose of PSI regarding WMD interdiction.

---

3   UNSCR 1540 (2004), S/RES/1540 (2004), April 28, 2004.

b. At their own initiative, or at the request and good cause shown by another state, to take action to board and search any vessel flying their flag in their internal waters or territorial seas, or areas beyond the territorial seas of any other state, that is reasonably suspected of transporting such cargoes to or from states or non-state actors of proliferation concern, and to seize such cargoes that are identified.

Section (b) asks countries to board their flagged vessels no matter where they are (including on the high seas) to interdict WMD cargoes. Since a vessel is considered to be the territory of the country whose flag it flies, no legal issues are associated with such an action.

c. To seriously consider providing consent under the appropriate circumstances to the boarding and searching of its own flag vessels by other states, and to the seizure of such WMD-related cargoes in such vessels that may be identified by such states.

Section (c) of Provision 4 differs from Section (b) in that countries are asked to consider granting other countries the permission to search their flagged vessels in order to interdict WMD cargoes. Again, since a vessel is considered to be the territory of the country whose flag it flies, no legal issues are associated with such an action if done with permission. To facilitate searches in such situations, the United States has entered into ship-boarding agreements with nine countries: Bahamas, Belize, Croatia, Cyprus, Liberia, Malta, Marshall Islands, Mongolia, and Panama. These agreements provide authority on a bilateral basis to board ships registered under the flag of one of these nine countries and believed to be carrying suspect cargoes. They also establish procedures for boarding and searching vessels in international waters. Under these agreements a vessel may be searched as little as two hours after a request has been made by a third party. The flags of these nine countries are often used by shippers as "flags of convenience," and a majority of the world's shipping operates under these flags: about 70 percent of total tonnage, comprising about 300 million gross registered tons.

d. To take appropriate actions to (1) stop and/or search in their internal waters, territorial seas, or contiguous zones (when declared) vessels that are reasonably suspected of carrying such cargoes to or from states or non-state actors of proliferation concern and to seize such cargoes that are identified; and (2) to enforce conditions on vessels entering or leaving their ports, internal waters or territorial seas that are reasonably suspected of carrying such cargoes, such as requiring that such vessels be subject to boarding, search, and seizure of such cargoes prior to entry.

This is the most controversial section of Provision 4. It calls on countries to interdict a vessel suspected of carrying WMD cargoes while it is in their internal waters, territorial seas, or contiguous zones when the vessel is not flying their flag and they lack permission from the flag country. When suspected vessels of this type are in a country's internal waters, interdictions are always permissible, because countries have total sovereignty over their internal waters. When such vessels are in a country's territorial seas or contiguous zones, however, the legal authority is suspect, because the UN Convention on the Law of the Sea (UNCLOS) grants vessels the "right of innocent passage" through territorial seas and contiguous zones.[4] Indeed, this point has led such countries as China to question PSI's legality. However, the introductory

---

[4]   *United Nations Convention on the Law of the Sea of 10 December 1982*, Section 3.

portion of Provision 4 states specifically that interdiction actions should only be undertaken "to the extent their national legal authorities permit and consistent with their obligations under international law and frameworks." In other words, PSI does not call for illegal action. But the circumstances under which vessels in states' territorial seas and contiguous zones can be interdicted when the country whose flag they fly does not grant permission are unclear.

> e. At their own initiative or upon the request and good cause shown by another state, to (a) require aircraft that are reasonably suspected of carrying such cargoes to or from states or non-state actors of proliferation concern and that are transiting their airspace to land for inspection and seize any such cargoes that are identified; and/or (b) deny aircraft reasonably suspected of carrying such cargoes transit rights through their airspace in advance of such flights.

Section (e) indicates that interdiction of aircraft flying over a country's territory might also take place. Such an interdiction would be legal under the authority of the Convention on International Civil Aviation,[5] which specifically states that contracting states are "entitled to require the landing at some designated airport of a civil aircraft flying above its territory without authority."[6]

> f. If their ports, airfields, or other facilities are used as transshipment points for shipment of such cargoes to or from states or non-state actors of proliferation concern, to inspect vessels, aircraft, or other modes of transport reasonably suspected of carrying such cargoes, and to seize such cargoes that are identified.

Section (f), Provision 4's final section, indicates that shipments over land might also be subject to interdiction. Countries are considered to have complete sovereignty over their land territory, so no legal issues would be raised by such an interdiction.

Readings for Session 2: PSI Design and Interdiction Principles

Christer Ahlstrom, "The Proliferation Security Initiative: International Law Aspects of the Statement of Interdiction Principles," Chapter 18, *SIPRI Yearbook 2005: Armaments, Disarmament and International Security*, United Kingdom: Oxford University Press, 2005. Downloadable as of February 28, 2009, at:
http://yearbook2005.sipri.org/ch18/ch18
This article provides a good background on the history of the interdiction of ships and the legal issues associated with such interdictions.

Michael Byers, "Policing the High Seas: The Proliferation Security Initiative," *The American Journal of International Law*, 98(3), July 2004, pp. 526–544.
This article takes a broad historical view of ship interdiction and how the law in this area has changed over time.

---

[5]   International Civil Aviation Organization, "Convention on International Civil Aviation," Doc 7300/9 (9th edition), 2006.

[6]   International Civil Aviation Organization, 2006, p. 3 (English: Article 3 *bis*, paragraph b).

Samuel E. Logan, "The Proliferation Security Initiative: Navigating the Legal Challenges," *Journal of Transnational Law & Policy*, 14(2), Spring 2005. Downloadable as of February 28, 2009, at:
http://www.law.fsu.edu/journals/transnational/backissues/issue14_2.html
This article examines the legal issues limiting PSI and discusses possible legal approaches to overcoming these limitations.

U.S. Laws Relevant to PSI

The United States has long identified the possession of weapons of mass destruction (WMD) by states and non-state actors "of proliferation concern" as a serious threat to domestic and international peace and security. It has been pursuing an international effort and a domestic effort to combat WMD proliferation. On the international side, it joins forces with countries of like mind for the endeavor. On the domestic side, it has established and is constantly improving a legal framework to control the export and import of illicit items.

The Proliferation Security Initiative (PSI) is an important element in the U.S. counter-proliferation effort. The most controversial activity of PSI is interdiction. Specifically, PSI's interdiction principles state that

> PSI participants are committed to the following interdiction principles to establish a more coordinated and effective basis through which to impede and stop shipments of WMD, delivery systems, and related materials flowing to and from states and non-state actors of proliferation concern, consistent with national legal authorities and relevant international law and frameworks, including the UN Security Council.[1]

Thus, the United States and other PSI participants are expected to conduct interdiction as well as other PSI activities consistent with their domestic laws and international commitments.

This session focuses on the U.S. laws and authorities that are pertinent to the execution of PSI activities, including interdiction. Session 4, International Agreements Relevant to PSI, covers the pertinent international treaties, agreements, and efforts. The first topic is the role of the U.S. military in supporting PSI. The objective is to indicate the scope and types of PSI activities in which Geographic Combat Command (GCC) personnel will be involved. The second topic is U.S. criminal laws and intelligence gathering as these pertain to the acquisition, transfer, and possession of WMD, their delivery systems, and related materials, which we refer to, for convenience, as *WMD items*. The third topic is export control of WMD items; the fourth is border control, which helps prevent not only the export, but also the import of WMD items.

---

[1] U.S. Department of State, Under Secretary for Arms Control and International Security, Bureau of International Security and Nonproliferation, *Interdiction Principles for the Proliferation Security Initiative*, Bureau of International Security and Nonproliferation, Washington, D.C., September 4, 2003.

U.S. Military Support to PSI

On March 1, 2007, the Chairman of the Joint Chiefs of Staff (CJCS) issued Instruction 3520.02A to set forth "policy and … procedures for the planning and execution of US military support to the PSI activity program."[2] The military agencies' responsibilities for the development of an effective WMD interdiction effort can be traced back to National Security Presidential Directive (NSPD) 17, *National Strategy to Combat Weapons of Mass Destruction*, December 2002, and NSPD 20, *Counterproliferation Interdiction*, dated November 2002. PSI is not an international or national organization and has no headquarters, chain of command, or assigned forces. Instead, a PSI Operational Experts Group (OEG) meets periodically, on behalf of all PSI participants, to guide PSI activities, including PSI's exercises, enhancement of participants' WMD interdiction capabilities, and the building of support for PSI.

The military is involved in two categories of PSI activities: exercises and interdiction operations. The first of these includes PSI-related training events. The Joint Staff, Office of the Secretary of Defense (OSD), GCCs,[3] and U.S. Strategic Command (USSTRATCOM) coordinate the participation of U.S. military forces in exercises and interdiction operations. GCCs are encouraged to incorporate PSI exercises into their existing exercise program, whether they are United States only, bilateral, or multilateral. USSTRATCOM serves as the supporting combat command for integration, synchronization, and execution of Department of Defense (DoD) efforts to combat WMD, and supports GCC in PSI exercise planning. GCCs are represented at the OEG meetings as required or as requested by OSD and the Joint Staff.

All PSI exercises and operations are conducted consistent with national legal authorities, relevant international law, and international or national organizations. With OSD and/or the Joint Staff providing policy guidance, GCCs serve as the lead for U.S.-hosted PSI exercises within their area of responsibility (AOR). They also participate in exercises within their AOR that are led by other PSI participants. The GCCs provide feedback and lessons learned to U.S. Joint Forces Command (USJFCOM), Joint Staff, and USSTRATCOM on exercise issues. They provide subject-matter experts (SMEs) and support to international PSI meetings.

Criminal Laws and Intelligence Gathering

The United States has laws in place to make it a crime, except under certain, very limited circumstances, for any individual to acquire, transfer, and possess WMD items.[4] Also proscribed are conspiracies, attempts, or threats to use WMD. The United States prosecutes hoax cases involving WMD, as these can seriously disrupt normal government or business operations and waste scarce resources. U.S. law prohibits teaching or demonstrating how to make or use a WMD. It is also a crime to provide material support or resources within the United States to

---

[2]   Chairman of the Joint Chiefs of Staff, *Proliferation Security Initiative (PSI) Activity Program*, CJCS Instruction 3520.02A, March 1, 2007 (current as of May 20, 2008).

[3]   Both terms—*Geographic Combat Commands* and *Regional Combat Commands*—appear in the literature. We use *Geographic Combat Commands*.

[4]   The information presented here draws from "United States Report to the Committee Established Pursuant to Resolution 1540 (2004): Efforts Regarding Security Council Resolution 1540," S/AC.44/2004/(02)/5, annex to letter dated October 12, 2004, to Chairman of UN Security Council Committee, October 14, 2004.

anyone intending to use the support or resources to commit terrorism-related crimes, including those involving WMD.

In an effort to share terrorism-related information among federal and state agencies, the Federal Bureau of Investigation (FBI) created the Joint Terrorism Task Forces. These are teams of state and local law enforcement officers, FBI agents, and other federal agents working together to investigate and prevent acts of terrorism, including those related to WMD. The FBI is also responsible for conducting and coordinating counterintelligence (CI) activities in the United States against intelligence and terrorist activities, including those pertaining to WMD items, that are conducted for foreign powers, organizations, or persons.

The U.S. Department of Homeland Security (DHS) provides analysis of terrorist threats, including WMD, to the United States and compares threats against vulnerabilities. It has established the Homeland Security Information Network to share all available information with those who need it.

Attorneys in the Counterterrorism Section of the U.S. Department of Justice's Criminal Division provide prosecutorial resources to WMD proliferation prevention and prosecution.

The criminal laws can have deterrence effects on individuals willing to help others acquire WMD items for financial gains. Monitoring these persons' activities may flag their shipments for further inspection and thus allow PSI participants to be more focused in their searches for and interdictions of illicit WMD items.

Export Control

Export of defense articles, including technical data and defense services, requires licenses pursuant to the Arms Export Control Act (AECA). Export and reexport of sensitive U.S.-origin dual-use items[5] and nuclear-related items also require licenses to be consistent with the Export Administration Act (EAA) of 1979, the Export Administration Regulations (EAR),[6] and the Atomic Energy Act of 1954. Any person violating any license requirement may be subject to civil fines. Those who willfully violate or willfully attempt to violate any license requirement may be subject to criminal penalties, including fines and/or imprisonment.

If the U.S. Secretary of State determines that a foreign person has contributed or attempted to contribute materially to the efforts of any foreign country or project of proliferation concern to acquire or produce WMD or missiles capable of delivering them, measures that can be taken include a ban on U.S. government procurement of goods, technology, or services from the designated foreign person; a ban on any U.S. government assistance to the designated foreign person; and a ban on importation into the United States from the designated foreign person.

The EAR prohibits export and reexport of any items to persons designated as terrorist entities by the U.S. Department of the Treasury. The Treasury department also compiles its "Entity List," which identifies specific end users in countries throughout the world that pose a proliferation concern.[7] For most of these end users, a license is required for all exports subject

---

5   *Dual-use items* are commercially available items that can be used or adapted for military use.

6   U.S. Department of Commerce, Bureau of Industry and Security, "Export Administration Regulations," Web site, last updated January 16, 2009.

7   U.S. Department of the Treasury, "Entity List," Supplement No. 4 to Part 744 of the Export Administration Regulations, December 5, 2008.

to the EAR. By watching the movements of these designated persons and the activities of these entities, and sharing intelligence about them, PSI participants have a better chance of uncovering their procurement of WMD items.

The U.S. Department of Energy controls exports of nuclear technology. It requires assurances from the recipient government that transferred U.S. technology or services will not be used for any military purpose and will not be retransferred to another country without prior U.S. government consent.

The U.S. Department of Commerce controls exports of dual-use items. Sensitive items are identified on the "Commerce Control List" as items that the United States considers of significant value to the development of WMD and other military programs of concern.[8] Certain items on the list may require a license for export to all destinations; others may be eligible for a license exception if the recipient country is a close ally or partner.

The United States also implements "catch-all controls" that require exporters to obtain a license to export any U.S.-origin item, even a non-controlled item, if they know or are informed that the item will be used by certain countries for prohibited WMD or missile programs.

An inconsistency between the shipper's export declaration and the bills of lading or other intelligence can make U.S. authorities suspect that a WMD item is on its way to a foreign country. This lead allows PSI participants to coordinate their efforts immediately for further investigation and even interdiction. PSI interdiction would be futile, like trying to find a needle in the haystack, in the absence of a clue.[9]

Border Control

The United States has two layers of defense in controlling imports: (1) import control at the U.S. border, and (2) control at foreign ports, which relies on cooperating exporting countries to prevent illicit shipments to the United States from leaving their ports. These countries have their own export control regulations and procedures and their border control to stop illegal merchandise from being exported. In January 2002, the United States initiated the Container Security Initiative (CSI) to help itself and other participating countries prevent the use of containerized shipping to conceal a WMD item.[10] The focus here is on control at the U.S. border.

The U.S. Department of Homeland Security (DHS) and its agencies have substantial domestic legal authority in border control to interdict and prevent the illegal import, export, or transit of illegal items, including WMD items, in the United States. The key agencies are Immigration and Customs Enforcement (ICE), Customs and Border Protection (CBP), the U.S. Coast Guard (USCG), and the Transportation Security Administration (TSA).

ICE is the largest investigative arm of DHS. Of its four law-enforcement divisions and several support divisions, the ones that can provide the most help to PSI are the Office of Intelligence and the Office of Investigations. The Office of Intelligence is responsible for collecting,

---

[8]   U.S. Department of Commerce, Bureau of Industry and Security, "The Commerce Control List," Part 774 of the Export Administration Regulations (EAR).

[9]   For a discussion of the difficulties of detecting WMD items without prior intelligence, see Session 6, Detection of WMD, Their Delivery Systems, and Related Materials.

[10]   CSI is discussed in Session 4, International Agreements Relevant to PSI.

analyzing, and sharing strategic and tactical intelligence data for use by the operational elements of ICE and DHS.[11] The intelligence data in this case are data related to the movement of people, money, and materials into, within, and out of the United States. Thus, whenever the data uncover a sender or recipient with a prior record of smuggling or the motivation to send or receive WMD items, this intelligence can help PSI decide where and when to conduct an interdiction. The Office of Investigations investigates a wide range of national security, financial, and smuggling violations, including illegal arms exports. Again, such information can lead to more fruitful PSI actions. ICE has more than 15,000 employees working within the United States and around the world. They work closely with foreign governments to perform their duty. Thus, for example, if ICE were to notice that a suspected shipment was on its way to a foreign country, it could issue a "redelivery order" to that country's government pursuant to a Customs Mutual Agreement request. The shipment would then be redelivered to the U.S. This example illustrates that PSI can stop the transfer of WMD items without relying on its most controversial activity, interdiction at sea.

ICE and CBP officers can search importing and exporting merchandise and cargo and can search persons nonintrusively at the border without a warrant. Moreover, the United States has numerous customs agreements and border inspection and pre-inspection arrangements with other countries. The United States has also signed more than 50 bilateral agreements of mutual legal assistance with other countries. These agreements provide mutual assistance in global investigations and prosecutions of criminal cases, including those involving WMD proliferation.

Readings for Session 3: U.S. Laws Relevant to PSI

Chairman of the Joint Chiefs of Staff, *Proliferation Security Initiative (PSI) Activity Program*, CJCS Instruction 3520.02A, March 1, 2007 (current as of May 20, 2008). Downloads as of February 19, 2009:
http://www.dtic.mil/cjcs_directives/cdata/unlimit/3520_02.pdf

"United States Report to the Committee Established Pursuant to Resolution 1540 (2004): Efforts Regarding Security Council Resolution 1540 (2004)," S/AC.44/2004/(02)/5, annex to letter dated October 12, 2004, to Chairman of UN Security Council Committee, October 14, 2004. Downloadable as of February 19, 2009, at:
http://www.un.org/Docs/journal/asp/ws.asp?m=S/AC.44/2004/(02)/5
Read parts pertaining to U.S. domestic laws and authorities.

For Further Study
National reports on the implementation of Security Council Resolution 1540, submitted to the Chairman of the Security Council Committee established pursuant to Resolution 1540 (2004). Downloadable by country as of February 13, 2009, at:
http://www.un.org/sc/1540/nationalreports.shtml
These reports describe how individual countries "adopt and enforce appropriate effective laws which prohibit any non-State actor to manufacture, acquire, possess, develop, trans-

---

[11]   U.S. Immigration and Customs Enforcement, "About: Offices Within ICE," last modified December 8, 2008.

port, transfer or use nuclear, chemical or biological weapons and their means of delivery, in particular for terrorist purposes" (UNSCR 1540, paragraph 2). From these reports, one can learn about the domestic laws of other PSI participants, as well as about those for current nonparticipants.

International Agreements Relevant to PSI

International agreements provide legitimacy to the Proliferation Security Initiative (PSI) by making nonproliferation a universal norm. Enhanced legitimacy induces countries to become PSI participants and mitigates the controversy associated with PSI interdictions.

The most controversial type of interdiction is interdiction of a suspected illicit shipment of weapons of mass destruction (WMD), their delivery systems, and related materials (which we refer to, for convenience, as *WMD items*) outside the territories of all PSI participants—whether at sea, on land, or in the air. This form of interdiction is the principal reason some countries are hesitant to join PSI. Fortunately, it is required only as a last resort, because PSI participants have a cooperative, *defense-in-depth* strategy they can pursue to prevent WMD items from falling into the hands of states and non-state actors of proliferation concern.

The first layer of the defense-in-depth strategy is to make it difficult for any nation or any person to acquire WMD items. If this defense fails, the shipment can be stopped before it leaves the port of export. Another layer of defense entails inspecting the shipment at a PSI participant's port, which is an option if the shipment makes a call at a participant's port en route elsewhere or if a participant's port is the shipment's final destination. The next defense is to interdict the carrier if it is flying the flag of a PSI participant; this can be done both inside and outside PSI participants' territories. Only when all of these layers of defense, or options, are infeasible do PSI participants consider using the most controversial type of interdiction—namely, interdicting a carrier flying a non-PSI-participant flag either inside or outside a PSI participant's territory.

The focus here is on how international agreements and programs, as well as domestic laws, help make this defense-in-depth strategy easier and more effective.[1] We have classified treaties, conventions, agreements, and programs—whether they are international, multinational, or bilateral—into seven groups, A through G:

- Group A arrangements have nonproliferation objectives similar to PSI. They help build international consensus that proliferation of WMD items is a serious threat to international peace and security.
- Group B focuses on terrorism. These agreements help prevent non-state actors from gaining access to WMD items
- Group C consists of United Nations Security Council Resolutions (UNSCRs). These help provide legal justification for PSI interdiction. When PSI participants interdict, they

---

[1]   This session focuses on international law and frameworks. National legal authorities are addressed in Session 3, U.S. Laws Relevant to PSI.

make sure that their actions are consistent with "national legal authorities and relevant international law and frameworks, including the UN Security Council."[2] Further, it is preferable for PSI interdictions to be supported by a UNSC resolution.

- Group D consists of bilateral ship-boarding agreements. These facilitate the efforts of U.S. and PSI partners to board ships suspected of carrying illicit WMD items.
- Group E is the UN Convention on the Law of the Sea (UNCLOS). This is the arrangement that is most often singled out as making countries hesitant to join PSI.
- Group F consists of U.S.-sponsored programs. These assist other countries in detecting concealed WMD items and establishing better export and import control.
- Group G covers all other arrangements relevant to PSI.

These seven groups are discussed in turn. In addition, we indicate whether the United States and the five "holdout" countries—Indonesia, Malaysia, India, Pakistan and China—have signed or ratified these agreements or joined these programs .[3]

Group A: Nonproliferation Treaties and Agreements

Group A arrangements can be further divided into four subgroups: Group A1 for nuclear weapons, Group A2 for chemical and biological weapons, Group A3 for missiles, and Group A4 for conventional arms and dual-use items.

Group A1 consists of nuclear treaties and agreements relevant to PSI:

- Nuclear Nonproliferation Treaty
- International Atomic Energy Agency (IAEA) Safeguard Agreement
- Partial Test Ban Treaty
- Comprehensive Nuclear Test Ban Treaty (as of February 2009, yet to enter into force)
- Convention on Physical Protection of Nuclear Material
- Nuclear Suppliers Group
- Zangger Committee.

Group A2 consists of conventions and agreements on chemical and biological weapons:

- Convention on the Prohibition of the Development, Production, Stockpiling and Use of Chemical Weapons and on Their Destruction (CWC)
- Convention on the Prohibition of the Development, Production and Stockpiling of Bacteriological (Biological) and Toxin Weapons and on Their Destruction (BTWC or BWC)
- BTWC Confidence Building Measures (CBMs)
- Australia Group (AG).

[2]   U.S. Department of State, Under Secretary for Arms Control and International Security, Bureau of International Security and Nonproliferation, *Interdiction Principles for the Proliferation Security Initiative*, Bureau of International Security and Nonproliferation, Washington, D.C., September 4, 2003.

[3]   The "holdout" countries are five key countries that have so far chosen not to affiliate with PSI. For further information on these five, their reasons for nonaffiliation, and details on which of these agreements have been signed by the five and the United States, see Charles Wolf, Jr., Brian G. Chow, and Gregory S. Jones, *Enhancement by Enlargement: The Proliferation Security Initiative*, MG-806-OSD, Santa Monica, Calif.: RAND, 2008.

Group A3 consists of control regime and code of conduct for missiles:

- Missile Technology Control Regime
- The Hague Code of Conduct Against Ballistic Missile Proliferation

Group A4 comprises an agreement on the transfer of conventional arms and dual-use items:

- Wassenaar Arrangement

These international treaties and agreements are building a universal norm against WMD proliferation, helping to justify or facilitate PSI activities in three ways: (1) More countries may welcome PSI interdiction as a major step forward in enforcement as they recognize that PSI has a purpose parallel with their existing international obligations in nonproliferation. (2) Such international agreements as the AG have a catch-all control similar to one in the U.S. domestic export control framework. State participants in these international agreements want exporters to notify the authorities if they are aware that nonlisted items are intended to contribute to proscribed activities. Thus, this catch-all provision allows the international community to cast a much wider net to catch transfers of illicit WMD items, making PSI interdiction easier and more effective. (3) The control lists and trigger lists in many of these international agreements provide specific WMD items on which PSI should focus its efforts to stem their transfer, again making PSI interdiction more effective.

Group B: Terrorism-Related Convention

The International Convention on the Suppression of Acts of Nuclear Terrorism, adopted in 2005 and entered into force in 2007, is a relatively recent convention.[4] There are about 12 other counterterrorism conventions and obligations covering terrorist bombings, financing, hostage taking, unlawful seizure of aircraft, violence to ships and their passengers, etc. Most of these arrangements were in force before the attacks on the United States on September 11, 2001, indicating that terrorism has long been a concern in the international community. These agreements provide moral, though not legal, support to PSI activities.

Group C: UN Resolutions

UNSCR 1540

The UN resolution most relevant to PSI is UNSCR 1540.[5] Originally, the United States wanted to use the UN to change international law and criminalize WMD proliferation activities in order to support PSI. China and Russia did not want that, however, and threatened to veto any resolution endorsing PSI.[6] Moreover, the final text was agreed to only after the United States accepted China's demand (accompanied by the threat of a veto) that a provision specifically

---

[4]   United Nations, "International Convention on the Suppression of Acts of Nuclear Terrorism," 2005.

[5]   UNSCR 1540 (2004), S/RES/1540 (2004), April 28, 2004.

[6]   William Hawkins, "Chinese Realpolitik and the Proliferation Security Initiative," February 18, 2005.

authorizing interdiction of vessels suspected of carrying WMD be dropped.[7] While UNSCR 1540 supports WMD/missile nonproliferation in general, it tends to focus on non-state actors and illicit private ventures involving WMD. It does not address WMD sales to nations, whereas PSI focuses on both states and non-state actors of proliferation concern.

UNSCR 1737

UNSCR 1737 aims to make Iran suspend all enrichment-related and reprocessing activities, including research and development, under verification by the IAEA.[8] For interdiction, especially on carriers not flying PSI-participant flags (such as ships) or in areas where PSI participants do not have indisputable jurisdiction, PSI pays close attention to UNSCRs such as 1737. UNSCR 1737 helps PSI in two ways: (1) It asks all states to take necessary measures, possibly including interdiction, to prevent Iran from obtaining the proscribed nuclear items. (2) Its annex lists names of entities and individuals designated by the Security Council as being engaged in, directly associated with, or providing support for Iran's proliferation-sensitive nuclear activities and for the development of nuclear weapon delivery systems. Thus, all PSI participants would monitor the movement of these designated individuals and alert the country whenever any one of them plans to make an entry into or exit from it. By closely tracking the individual and his/her contacts and inspecting his/her baggage, this individual's effort in procuring an illicit WMD item can be uncovered and stopped at the preferred inspection spots (i.e., at a PSI participant's exit or entry port), as described above for the defense-in-depth strategy. In fact, even if the shipment slips through customs, UNSCR 1737 can still provide justification for any state to interdict the carrier anywhere, provided that the individual is suspected of carrying items proscribed by UNSCR 1737, which states that "all [flag] States shall take the necessary measures."

UNSCRs 1803 and 1718

On March 3, 2008, the Security Council passed UNSCR 1803.[9] It, as well as UNSCRs 1696 (July 31, 2006) and 1747 (March 24, 2007), reinforces the commitment and effort to prevent Iran from developing a nuclear weapon capability, further helping PSI to stem the nuclear flow into Iran.

UNSCR 1718 serves a purpose for PSI against North Korea that is similar to the purpose that UNSCR 1737 serves for PSI against Iran.[10] Also, earlier resolutions, UNSCRs 1695 (July 15, 2006) and 825 (May 11, 1993), reinforce the commitment and effort against North Korean WMD and missile programs, further justifying PSI interdiction of WMD items to and from North Korea.

---

[7]   Mark Valencia, *The Proliferation Security Initiative: Making Waves in Asia*, Adelphi Paper 376, International Strategic Studies Institute, 2005, p. 48.

[8]   UNSCR 1737 (2006), S/RES/1737 (2006), December 27, 2006 (reissue of December 23 version).

[9]   UNSCR 1803 (2008), S/RES/1803 (2008), March 3, 2008.

[10]   UNSCR 1718 (2006), S/RES/1718 (2006), October 14, 2006.

Group D: Bilateral Ship-Boarding Agreement

Modeling its arrangements in the counternarcotics arena, the United States has entered into bilateral ship-boarding agreements with Bahamas, Belize, Croatia, Cyprus, Liberia, Malta, Marshall Islands, Mongolia, and Panama.[11] These agreements allow authorities, on a bilateral basis, to board sea vessels suspected of carrying illicit WMD items: "Either one of the parties to this agreement can request of the other to confirm the nationality of the ship in question and, if needed, authorize the boarding, search, and possible detention of the vessel and its cargo." Thus, these agreements facilitate, or even make possible, PSI interdiction.

Group E: Law of the Sea

Countries concerned about the legality of PSI cite the UN Convention on the Law of the Sea (UNCLOS) most often as an issue. For example, Indonesia, one of the five holdout countries, has expressed its view that PSI commitment would violate UNCLOS's stipulation of the protected right of innocent passage. Innocent passage and PSI interdiction can co-exist, however, and this issue should not be an obstacle for countries considering whether to affiliate with PSI.[12]

In May 2007, President Bush said that ratification of UNCLOS will secure U.S. sovereign rights over extensive marine areas, including the valuable natural resources they contain. Both Department of State and DoD officials have also been pushing for UNCLOS's ratification. A spokesperson for the U.S. Navy said that the Navy does support the treaty. On October 30, 2007, the Senate Foreign Relations Committee voted 17 to 4 to recommend ratification of UNCLOS and referred it to the full Senate for a vote, which, as of February 2009, had yet to occur.

Group F: Assistance Program

The Export Control and Related Border Security Assistance (EXBS) Program is the U.S. government's premier initiative to help other countries improve their export control systems. For example, EXBS funds Commodity Identification Training (CIT), whose curriculum is to educate and train customs inspectors and border enforcement personnel from around the world in techniques of detection and interdiction for the purpose of preventing illicit trade in items and technologies needed to manufacture WMD.

Given that many states, including the five holdout countries,[13] are improving their export and import control, especially since the terrorist attacks on the United States on September 11, 2001, this program is of value to them. Although countries do not need to join PSI to participate in EXBS, EXBS can help PSI in two ways. First, interaction with the United States

---

[11] U.S. Department of State, Under Secretary for Arms Control and International Security, Bureau of International Security and Nonproliferation, "Ship Boarding Agreements," Web page, with links to U.S.-country agreements (e.g., "Proliferation Security Initiative Ship Boarding Agreement with Belize").

[12] The coexistence of the right of passage and PSI interdiction is discussed in Wolf, Chow, and Jones, op. cit.

[13] Wolf, Chow, and Jones, 2008.

through EXBS should help countries to better understand U.S. intentions in combating pro-liferation and may resolve some concerns about joining PSI. Second, as EXBS helps improve these countries' import and export control frameworks and national inspection capabilities, the contribution they could make to PSI would be greater as PSI participants.

Group G: Other Agreements

The Container Security Initiative (CSI) was launched in 2002 by the U.S. Bureau of Customs and Border Protection (CBP), Department of Homeland Security. Its purpose is to increase security for container cargo being shipped to the United States. The screening of containers that pose a risk of terrorism is performed by the host nation at its ports that participate in CSI. U.S. CBP officers can station at these foreign ports to observe the inspection. CSI offers par-ticipating countries the reciprocal opportunity to enhance their incoming shipment security by sending their customs officers to major U.S. ports to target ocean-going, containerized cargo being exported from the United States to their countries. Clearly, the activities of CSI and PSI reinforce each other for stemming the flow of illicit WMD items. Also, the two benefits for participating in EXBS hold for those nations in joining CSI.

Other initiatives, such as the Security Freight Initiative and the Global Initiative to Combat Nuclear Terrorism, have a purpose similar to that of CSI and provide benefits to par-ticipants that are similar to those of CSI.

Readings for Session 4: U.S. Laws Relevant to PSI

"United States Report to the Committee Established Pursuant to Resolution 1540 (2004): Efforts Regarding Security Council Resolution 1540 (2004)," S/AC.44/2004/(02)/5, annex to letter dated October 12, 2004, to Chairman of UN Security Council Committee, October 14, 2004. Downloadable as of February 19, 2009, at:
http://www.un.org/Docs/journal/asp/ws.asp?m=S/AC.44/2004/(02)/5
Read parts pertaining to U.S. international commitments and obligations.

Center for Nonproliferation Studies, "Inventory of International Nonproliferation Organiza-tions and Regimes," undated. As of February 20, 2009:
http://www.cns.miis.edu/inventory/index.htm
This public reference, which is updated regularly, describes international and functional orga-nizations and regimes, international treaties, and membership of selected states.

National reports on the implementation of Security Council Resolution 1540, submitted to the Chairman of the Security Council Committee established pursuant to resolution 1540 (2004). Downloadable by country as of February 13, 2009, at:
http://www.un.org/sc/1540/nationalreports.shtml
These reports are equivalent to "United States Report to the Committee Established Pursu-ant to Resolution 1540 (2004): Efforts Regarding Security Council Resolution 1540 (2004)," (described above). They describe how individual countries "adopt and enforce appropriate

effective laws which prohibit any non-State actor to manufacture, acquire, possess, develop, transport, transfer or use nuclear, chemical or biological weapons and their means of delivery, in particular for terrorist purposes."[14]

[14] UNSCR 1540 (2004), 2004.

Incentives and Disincentives for PSI Participation

In addressing the issue of the incentives and disincentives—or pros and cons, benefits and costs—associated with PSI participation, we consider the perspectives of countries that have affiliated with PSI, as well as those key countries that have not—that is, the five holdout countries: China, India, Pakistan, Indonesia, and Malaysia.[1] We begin by discussing a useful distinction, that between the collective benefits associated with PSI activities and the particular benefits that accrue to each PSI participant.

Collective Benefits of PSI

Collective, or public, benefits are what economists construe as "public goods"—for example, clean air, well-paved roads, and law and order. Benefits of this sort are construed as public because they have two essential characteristics: (1) they are nonexclusionary—that is, available to all regardless of whether individuals or individual countries contribute to them[2]; (2) they are nonrivalrous—that is, addition of one or more individual or individual country to the beneficiaries does not diminish the benefits accruing to other individuals and individual countries.[3]

To the extent that countries (or at least most countries) prefer a less-proliferated to a more-proliferated world, and to the extent that PSI's activities deter proliferation, PSI "produces" a public good. To the extent that the deterrent effect of PSI's activities confers benefits on all countries whether or not they participate in PSI (i.e., the nonexclusionary characteristic of public goods), PSI confronts the classical "free-rider" problem—namely, that a country can realize at least part of the benefit of PSI affiliation without affiliating with PSI.

To overcome the free-rider problem, there must be specific benefits that accrue solely to those countries that formally endorse PSI[4] and participate in PSI activities. In other words, there must be benefits that do not accrue to free riders, or countries that simply stand by.

---

1 For a full discussion of this issue, see Charles Wolf, Jr., Brian G. Chow, and Gregory S. Jones, *Enhancement by Enlargement: The Proliferation Security Initiative*, MG-806-OSD, Santa Monica, Calif.: RAND, 2008.

2 To express this more rigorously, the nonexclusionary point essentially implies that the costs of excluding noncontributors to a public good are exorbitant relative to the benefits, because the benefits are not vested with property rights.

3 See Charles Wolf, Jr., *Markets or Governments: Choosing Between Imperfect Alternatives*, Chapter 2, "Market Failure," N-2505-SF, 1986.

4 See Session 2, PSI Design and Interdiction Principles.

Particular Benefits of PSI Affiliation

Affiliation with PSI offers countries particular, or national, benefits. One qualitative benefit that accrues to an affiliated country is more-cooperative strategic relations between itself and other PSI affiliates. For example, Saudi Arabia and Singapore may see endorsement of PSI as positively affecting their overall strategic relations with the United States. However, the five holdout countries—China, India, Pakistan, Indonesia, and Malaysia—are likely to consider this benefit outweighed by the possible costs of closer association with certain PSI participants, including the United States.[5]

Depending on each country's situation and calculus of its own national interests, it may also consider other aspects of PSI endorsement as potential benefits. For example, PSI participation includes workshops, training, and technical assistance to help countries improve import and export controls. In these and other respects, PSI affiliation facilitates and reinforces the benefits of more-rigorous control over the imports and exports of WMD items resulting from membership in the Container Security Initiative (CSI), Security Freight Initiative (SFI), the Global Initiative to Combat Nuclear Terrorism, and the Export Control and Related Border Security Assistance (EXBS) Program.[6]

Other benefits that may accrue directly to countries that participate in PSI include improved customs procedures, inspection capabilities, and shared intelligence among PSI members. As a consequence, vehicles carrying the flag of PSI participants are more likely to be "clean" and "safe" than are vehicles carrying the flags of nonparticipants. Hence, the risk exposure of PSI's nonparticipants—in the form, for example, of accidents in transit, or interdiction or interruption and hence delay in transit—may be calculably greater than the risk exposure of PSI participants. In effect, the transport of WMD systems or components involves increased risks to the points of origin, destination, and thoroughfare. Thus, PSI affiliation carries with it a presumptive reduction of these risks and hence a national benefit from PSI affiliation.

Disincentives/Costs Associated with PSI Affiliation

That there are disincentives and imputed costs associated with PSI endorsement is evident from the fact that several key countries—the holdout countries—view PSI affiliation as disadvantageous to them. These disincentives/costs can be categorized as stemming from

1. concerns about compromising national sovereignty and the independence of a nation's foreign policy
2. concerns about infringement of international law, particularly UNCLOS
3. internal political circumstances
4. misunderstandings about what PSI affiliation means in terms of commitments and volunteerism.

---

[5]   These five countries are the focus of Wolf, Chow, and Jones, 2008.

[6]   See Wolf, Chow, and Jones, 2008, pp. 53–55.

With respect to concerns about national sovereignty and an independent foreign policy, an Indonesian observer expressed what he saw as the drawbacks of PSI endorsement in the following terms:

> Initiation of interdiction . . . [of] suspected national flag vessels in international or national territorial waters . . . would potentially interfere with Indonesia's territorial sovereignty . . . [by] internationalization of Indonesian territorial waters and opening space within Indonesia's territory for external powers in their pursuit of WMD and other sensitive materials and technology.[7]

Furthermore, in some cases (for example, Indonesia and, to a lesser extent, Malaysia), maintaining the independence of a country's foreign policy is interpreted to mean "nonalignment with" and "independence of" the United States. This interpretation entails particular circumspection in undertaking or appearing to undertake obligations that might compromise independence as a result of an excessively close link to the United States. To the extent that PSI was at its inception and is still perceived to be "led" by the United States, and hence to be an adjunct of U.S. foreign and defense policy, affiliation with PSI may be viewed as entailing unwelcome costs.

Both China and Indonesia have expressed reservations about PSI on the grounds that participation entails commitments that would infringe on the right of innocent passage in international waters and violate the law of the sea. Countries that are especially concerned about this often make a clear distinction between transport of WMD items by non-state actors versus sovereign states. Their objections focus on transport by flagged vessels of states, and their acknowledgment that non-state actors might dissemble their non-state credentials in invoicing the cargo they transmit on flagged ships of sovereign states is somewhat ambiguous.

Disincentives in the third category stem from the political dynamics in democratic states, such as India and Indonesia. In these instances, the preservation of domestic political coalitions that underpin governmental support may be deemed at risk if the government formally endorses an enterprise that is viewed domestically as heavily weighted by U.S. influence.

In the fourth category are fundamental misunderstandings about the obligations formally associated with PSI affiliation. To be sure, PSI focuses on interdiction as a key component of global counterproliferation strategy. PSI also provides participating countries with opportunities to improve their interdiction-conducting capabilities and authorities. Furthermore, through PSI's Operational Experts Group (OEG), PSI develops operational concepts, organizes interdiction exercise programs, shares information about national legal authorities, and provides a basis for cooperation—including intelligence sharing—among partners on specific actions to be taken when circumstances warrant. However, PSI is an activity, not a formal treaty-based organization; and it does not obligate participating states to take any specific actions. Hence, a state's decision to participate in an interdiction is completely voluntary.

Failure to understand this quintessential characteristic of PSI affiliation—its voluntary nature—may, to a considerable extent, account for many of the disincentives associated with PSI affiliation. That said, however, PSI may be construed as creating a norm of "commitment" to and implementation of PSI's interdiction principles, which some countries may be reluctant

---

[7]   Memorandum to one of the authors from the Indonesian Center for Strategic and International Studies in Jakarta, April 2007.

to adopt. Though this commitment does not have the force of law, countries desiring to be seen as "normal" and hence compliant with international norms (China, for example) may consider PSI affiliation a powerful curb on their behavior that they might not want to accept.

Readings for Session 5: Incentives and Disincentives for PSI Participation

U.S. Department of State, Under Secretary for Arms Control and International Security, Bureau of International Security and Nonproliferation, *Proliferation Security Initiative Frequently Asked Questions (FAQ)*, Fact Sheet, May 22, 2008. As of February 25, 2009: http://www.state.gov/t/isn/115491.htm

Yann-hui Song, "An Overview of Regional Responses in the Asia-Pacific to the PSI," in *Countering the Spread of Weapons of Mass Destruction: The Role of the Proliferation Security Initiative,*" Pacific Forum CSIS's *Issues & Insights*, 4(5), July 2004, pp. 7–31.

Natalie Ronzitti, "The Law of the Sea and the Use of Force Against Terrorist Activities," in Natalie Ronzitti (ed.), *Maritime Terrorism and International Law*, Netherlands: Kluwer Law International, 1990, pp. 1–15.

Thom Shanker, "US Remains Leader in Global Arms Sales, Report Says," *New York Times*, September 25, 2003, p. A-12.

Josif Jofi, "The Proliferation Security Initiative: Can Interdiction Stop Proliferation?" *Arms Control Today*, June 2004. As of February 21, 2009: http://www.armscontrol.org/act/2004_06/Joseph

Mohamed ElBaradei, "7 Steps for Preventing Nuclear Proliferation," February 15, 2005. As of February 21, 2009: http://www.asahi.com/english/opinion/TKY200502150114.html

Detection of WMD, Delivery Systems, and Related Materials

The focus here is on the detection of illicit weapons of mass destruction (WMD), their delivery systems, and related materials that are being carried on a ship. For purposes of discussion, we first separate these items, which we refer to collectively as *WMD items*, into two groups: those that are radioactive and those that are nonradioactive.

Because the intelligence about suspected ships and cargoes can be faulty, it is a great advantage to be able to detect illicit WMD items without boarding a ship. Moreover, a PSI inspection team would want to avoid boarding the ship in order to minimize the disruption to trade. For nonradioactive WMD items, detection via a hovering aerial vehicle or a ship side by side with the suspected ship is highly unlikely to be successful for two reasons. First, if there is no intelligence cuing the inspection of a specific ship, there are simply too many ships to inspect. Second, even with intelligence zeroing in on a suspected ship, it is still hard to detect such items without boarding. The illicit WMD item can be concealed under a cover or in a container to prevent visual identification.[1] Additionally, a nonradioactive WMD item does not emit a tale-telling signal that can be detected by nonintrusive equipment from an inspection ship.

In the case of a radioactive WMD item, detection is at least theoretically possible without boarding if done at a close standoff distance. A radiation detector can detect radioisotopes that are not shielded. If the radioisotopes are heavily shielded, one approach is to try to detect the shielding instead, thereby raising the possibility that the intelligence was right and that a WMD item is on board. An X- or gamma-ray imaging machine can be used to detect a shielding that might be present to block detection of the radiation being emitted by the illicit item. Unfortunately, the devices for detecting radiation and shielding even from a close standoff distance (without boarding the ship) would be expensive and bulky. Worse yet, these devices would be unable to detect nonradioactive items, which include not only chemical and biological weapons, missiles, and their components, but also the nonradioactive components of radiological and nuclear weapons. In the near future and without a breakthrough in radiation detection and imaging technologies, PSI members cannot expect their inspection ships to be equipped with devices that can detect nonradioactive items and shielded radioactive items without being taken onboard the suspected ship.

With that said, we focus on a specific scenario: an onboard detection of WMD items once a PSI participant has been alerted by intelligence that a ship is suspected of carrying illicit WMD items. The U.S. Navy and U.S. Coast Guard both handle ship boarding that involves

---

[1]   A standard cargo container is 8 by 8 feet and 20 to 48 feet long. Even after a container is opened, a further search may be necessary to find a WMD item(s) hidden among legitimate items.

U.S. participants.[2] We begin with the feasibility of detecting illicit WMD items without opening a container or removing a cover in the case of radiological or nuclear weapons, their components, and their materials; chemical and biological items; and finally missile items. We then discuss the best place for conducting an onboard inspection and how the inspection can be helped to detect, locate, and identify a WMD item.

Physical Detection of Radiological and Nuclear Weapons and Their Materials

Such radiological materials as the industrial radioisotopes Co^{60}, Cs^{137}, and Am^{241}, which could be used to make dirty bombs, are relatively easy to detect nonintrusively at a close distance.[3] A hand-held passive radiation detector would be adequate for detecting these radioisotopes in many circumstances, but not when they are heavily shielded. However, such illicit items as a radioactive weapon without the radioactive material render a radiation detector useless. There is no available hand-held or mobile device that can detect such a nonradioactive item.

It is relatively much more difficult to detect a nuclear weapon or its fissile materials, such as plutonium (Pu) or highly enriched uranium (HEU),[4] as these materials are radioactively much weaker than Co^{60}, Cs^{137}, and Am^{241}. The current generation of (low-resolution) sodium iodide (NaI) and (intermediate-resolution) cadmium-zinc-tellurium (CZT) detectors may not have the energy resolution and/or sensitivity needed to identify the plutonium, so they may create an unacceptable number of false positives or be unable to distinguish plutonium from legitimate radioisotopes. With detectors this poor, a proliferator could hide a nuclear weapon or its fissile materials within legitimate radioisotopes for industrial and medical purposes. Fortunately, a more expensive, portable, cryogenic-cooled, passive, high-purity Germanium detector has the energy resolution and sensitivity needed to detect the gamma ray resulting from the plutonium's spontaneous fission, provided that it is not heavily shielded. Alternatively, the classic Helium-3 gas proportional counter can be used to detect neutrons resulting from the same spontaneous fission—again, provided that the plutonium is not heavily shielded. Unfortunately, an effective heavy radiation shielding can be provided by a lead layer of several centimeters (to attenuate the gamma ray) and a water or polyethylene or paraffin layer of several tens of centimeters (to attenuate the neutrons). Worse yet, such legitimate items as engine parts, shampoos, and fruits and vegetables (filled with water) may suffice for an effective shielding. Thus, it would be difficult for an inspection team to use hand-held or mobile equipment, without opening the containers, to detect and locate a radioactive item that is hidden and shielded. Further, the many containers on a ship are closely packed and stacked, so a boarding party would be unable to place a hand-held detector next to every container for more-effective detection.

---

[2]   U.S. air and land interdictions are handled by law enforcement and/or CBP personnel.

[3]   *Nonintrusive detection* is defined as detection of an item without opening the container in which or lifting a cover under which the item is hidden. Moreover, there are two classes of nonintrusive detection: (1) *passive* detection, in which radiation, vapors, or other releases from the item are detected, and (2) *active* detection, in which the item is irradiated or interacted with so that it releases radiation, vapors, or something else that is detectable.

[4]   Plutonium of any isotopic composition can be used in a nuclear weapon. HEU has a greater than 20 percent concentration of U^{235} or U^{233}. Although a 20 percent concentration, known as *weapons-usable uranium*, is adequate for making a crude, inefficient nuclear weapon, a typical nuclear weapon uses 85 percent or more of U^{235}, which is known as *weapons-grade uranium*.

Uranium's relatively low level of radioactivity makes HEU even harder to detect than plutonium, whether the detector is NaI, CZT, Germanium, or Helium-3.[5] This is particularly true for HEU that is not derived from reprocessed uranium, since this HEU does not contain any of the more radioactive and detectable radioisotopes, such as U^{232}, that would be present in HEU derived from or contaminated with uranium from reprocessed reactor fuel. Moreover, if uranium is shielded, it is very likely to escape detection by passive radiation monitors.

To detect fissile, fissionable, or special nuclear material (SNM), a pulsed fast neutron analysis (PFNA) inspection system would be better.[6] It generates high-energy neutrons to fission the material, including HEU without reprocessed uranium, and to effect the emission of gamma rays for detection and identification. This still nonintrusive method can also be used to detect an explosive by causing gamma-ray emissions from its high nitrogen and oxygen contents. The neutrons from the inspection system can penetrate deep into full cargo containers to detect nuclear material and explosives, except when the cargo contains materials, such as water, that can absorb or stop the neutrons before they reach the item. In any case, the inspection system needs a large radiation-shielded building in which to conduct the cargo scans, and the cargo needs to be moved to and from the scanning facility. In addition, the system should be accompanied by a gamma-radiographic or high-energy X-ray detection system to detect the presence of shielding. These inspection systems are much more effective in port than in a ship at sea,[7] which means that the inspection team would want to escort the seized ship to a port before conducting the inspection.

Physical Detection of Chemical and Biological Weapons or Agents

For chemical warfare agents and toxic industrial chemicals, hand-held chemical detectors, which aim to nonintrusively detect these chemicals' vapors, can be used. However, this method relies on leakage of the chemical agent from its own container into the cargo container and surrounding air, and a proliferator that can gain access to chemical agents is likely to be able to access to a leak-proof chemical container, as well. One possibility is the PFNA inspection system, discussed above, which can nonintrusively detect some chemical agents, such as Sarin, by measuring the gamma rays generated by neutron irradiation. These gamma rays can easily penetrate the leak-proof container and be detected. However, as discussed above, a PFNA inspection would be much more practical and efficient if performed in port rather than at sea.

---

[5]   For example, the decay rate (or rate of emission of radioactivity) for Pu^{239} is hundreds of times less than that for Cs^{137}, whereas the decay rate for U^{235} is 30,000 times lower than that for Pu^{239}. Nuclear Threat Initiative, *A Tutorial on Nuclear Weapons and Nuclear-Explosive Materials—Part Five*, July 2005.

[6]   Fissionable materials (such as U^{238}) are composed of nuclides for which fission with neutron is possible. Fissile materials (such as U^{235}, U^{233}, Pu^{239}, and Pu^{241}) are fissionable by thermal (slow-moving) neutrons. Some authorities restrict the term fissionable materials to non-fissile materials or to materials fissionable by fast, but not slow, neutrons. SNM is defined as plutonium, U^{233}, or uranium enriched in the isotopes U^{233} or U^{235} in *The Atomic Energy Act of 1954* (as amended), Public Law 83-703, 68 Stat. 919, August 30, 1954, Title I: Atomic Energy.

[7]   While there are portable X-ray imaging systems available, they are not effective for searching inside unopened cargo containers, because the containers are too large and too numerous. Moreover, containers are closely packed in a ship, which means the X-ray system cannot be placed close to every container for higher-resolution imaging.

As in the chemical case, a smuggler who knows where to acquire bio-weapon agents should be able to access a leak-proof biological container, as well. Moreover, biological agents are not volatile. Even if their containers develop a leak, detection is much more difficult than in the case of chemical vapors.

In sum, at sea, nonintrusive detection of chemical or biological agents is much more difficult than nonintrusive detection of radioactive items. In port, however, the disparity is much less when detectors such as PFNA are used.

Physical Detection of Missiles and Their Components

A full missile or even its stages are bulky. It may be possible for an inspection team to discover the presence of such items by looking under the cover after boarding a ship. However, if the items are small missile components, such as accelerometers, or missile materials, such as oxidizer substances, detection at sea would be difficult without opening the cargo container for a physical search.

Where and How to Conduct an Inspection of WMD Items

We start this section with a reclassification of the WMD items into three categories, based on the discussion above, so as to discuss how easy or difficult it will be for the inspection team to carry out an inspection on board a ship. Then, we address the issues of where and how to conduct a PSI interdiction on shipborne cargoes.

Reclassification of WMD Items

WMD items can be fully assembled chemical, biological, radiological, and nuclear weapons and missiles; their agents, materials and fuels; their parts and components; and the production equipment for these weapons and their parts. In addition to radioactive isotopes for dirty bombs, fissile materials for nuclear bombs, and explosives used in radiological and nuclear bombs, there are numerous WMD items that call for license to export or are restricted for export in the Australia Group's (AG's) "Australia Group Common Control Lists"[8] and the lists in the Missile Technology Control Regime's (MTCR's) *Equipment, Software and Technology Annex.*[9] The AG lists are the common export control lists for chemical weapon precursors, biological weapon agents, plant and animal pathogens, and dual-use chemical and biological production equipment. AG participants require licenses for the export of these dual-use items.[10] The MTCR aims "to restrict the proliferation of missiles, complete rocket systems, unmanned air vehicles, and related technology for those systems capable of carrying a 500 kilogram payload at least 300 kilometers, as well as systems intended for the delivery of weapons of mass destruction (WMD)."[11]

---

[8]   Australia Group, "Australia Group Common Control Lists," © 2007.

[9]   Missile Technology Control Regime, *Equipment, Software and Technology Annex*, updated November 5, 2008.

[10]   *Dual-use items* are commercially available items that can be used or adapted for military use.

[11]   Missile Technology Control Regime, "Objectives of the MTCR," undated.

Our reclassification of WMD items into three categories is as follows.

Category 1: WMD Items Detectable Even When Concealed. This category contains radio-active and some nonradioactive WMD items that can be detected with available sensors and means even when concealed inside a shipping container. However, as discussed above, these items are much more likely to be detected in port than at sea, because the better detectors are bulky and require more space both to operate and so that cargoes can be moved individually to the scanning facility. Also, these detectors are more expensive, making them more efficient if kept at centralized and easily accessible sites such as ports (as opposed to on ships). Examples of items in Category 1 are as follows:

- plutonium
- highly enriched uranium (HEU)
- assembled nuclear weapon
- low enriched uranium
- conventional explosives (a WMD weapon may contain explosives).

Category 2: Bulky WMD Items. This category contains items that an inspection team can identify visually and/or with a simple test and/or after consultation with SMEs via electronic exchange of pictures of and data on the suspected items soon after the team has boarded the ship and the items have been exposed. Of course, these items can also be detected, located, and identified in port. Examples of items in Category 2 are as follows:

- missile stage
- missile liquid fuel
- missile solid fuel
- centrifuges.

Category 3: Smaller WMD Items Not Detectable If Concealed. This category contains smaller WMD items that, if concealed, cannot be detected, which means the difficulty lies in finding them in the first place. In the rare instances in which specific intelligence leads to their location, some may be identified after consulting a reference manual full of pictures and data[12] and/or consulting with SMEs electronically, as discussed for Category 2 items. However, some might be so similar in appearance to legitimate items that even SMEs cannot distinguish them. In most cases, items in Category 3 cannot be detected at sea. A thorough search in port would be necessary to assure the inspection team that illicit WMD items are not hidden in one or more of the containers. Examples of Category 3 items are as follows:

- components (such as accelerometers) and materials (such as oxidizer substances) not already listed in Categories 1 and 2, and test and production equipment for missile components and materials, as described in the MTCR lists
- assembled chemical and biological weapons
- anthrax in a sprayer
- ready-to-go chemical and biological weapon agents

---

[12] For example, the 2007 WMD and missile reference manual produced by Sandia National Laboratories.

- chemical and biological weapon precursors and agents, chemical and biological agent or weapon production equipment (such as reaction vessels, agitators, condensers, distillation or absorption columns, incinerators, fermenters, and centrifugal separators), and plant and animal pathogens as described in the AG lists.

Military Personnel as Inspectors

Military personnel serving as inspectors are also faced with problems other than their inability to identify many WMD items. They do not have the subject-matter expertise and skill in conducting various tests needed to confirm that the items found are indeed illicit WMD items. Finally, inspectors may need to wear particular protective gear and follow specific procedures in order to safely conduct a given detection or identification test.

Given the great variety of tests for different WMD items, military personnel who are serving as inspectors simply may not be able to perform many of the tests. These problems point to the importance of *reach-back*, or real-time or near-real-time communications with SMEs who can help validate or invalidate a WMD item by talking to inspectors who are on board the ship and/or seeing pictures and/or data about the suspected item. A positive identification is a clear justification that the ship should be subject to a thorough inspection, which is most effectively conducted in a port. Unfortunately, many WMD items cannot be detected in the first place by the boarding party, because they are shielded or simply hidden in a few of the many containers. In these situations, which are the much more common type, reach-back cannot help. Again, in-port search is the only way to find out.

Measures That Can Help Detection of Illicit WMD Items

In January 2002, the Commissioner of Customs and Border Protection (CBP) announced the Container Security Initiative (CSI), whose objective is to identify maritime containers that pose risk of terrorism at foreign ports before they are shipped to the United States.[13] There are 58 foreign ports participating in CSI, accounting for 85 percent of U.S.-bound container traffic.[14]

CSI helps detect illicit WMD items by presenting obstacles and risks to proliferators when they attempt to transfer such items through CSI ports. Further, CSI allows PSI members to focus their efforts on transfers involving ports that do not participate in CSI. There are also other, similar programs that are complementary to PSI in preventing WMD items from falling into the hands of state and non-state parties of proliferation concern. For example, the Second Line of Defense Program aims to prevent illicit trafficking in nuclear and radiological materials by securing international land borders, seaports, and airports that may be used as smuggling routes for these materials. The Megaports Initiative supplements the CSI by focusing on SNM and radioactive material suitable for nuclear and radiological weapons.[15] Also making illicit transfer of WMD items more difficult are the Community Regime for the Control of Exports of Dual-Use Items and Technology (established by the Council of the European Union Regu-

---

[13] See Session 4, International Agreements Relevant to PSI.

[14] U.S. Department of Homeland Security, "Container Security Initiative Ports," Web page, last reviewed/modified October 20, 2008.

[15] See National Nuclear Security Administration, "Megaports Initiative," undated.

lation No. 1334/2000 on June 22, 2000);[16] and Export Controls under United Nations Security Council Resolution (UNSCR) 1540 (2000) and 1673 (2006).[17]

Tamper-indicating devices, such as seals, can be used to indicate whether a container has been tampered with. For example, a smuggler might try to open a legitimate container ready for shipping in order to insert an illicit WMD item into it. There are increased efforts to require clear declaration of contents and other information before a cargo's arrival at the port. Moreover, personnel who have access to the cargo must clear a background check and can access containers only at places where their actions are closely watched and monitored. Also, it is important that countries use physical protection, material control, and material accounting to protect and monitor their WMD items so that such items are not stolen and later transferred to proliferators. When PSI performs its activities in conjunction with all these actions and countermeasures, the effectiveness in meeting PSI's goal of detecting and preventing the transfer of illicit WMD items can be greatly improved.

Where and How to Conduct an Inspection
As is evident from the previous discussion, it is ineffective for PSI members to randomly search ships for illicit WMD items. Members need to share intelligence pertaining to suspicious persons and shipments. This intelligence can come from a close watch of the activities of individuals and entities designated by the UN Security Council or individual PSI members as engaging in or associated with illicit programs of countries of concern. The intelligence can also come from exporters that suspect an item is intended for illicit WMD purposes and report their suspicions to authorities under the domestic or international catch-all control provision.[18] Exporters, and others, can also report possible violations of the Export Administration Regulations (EAR) to the U.S. Department of Commerce.[19] The intelligence can come from an inconsistency between a shipper's export declaration and the bills of lading. However, proliferators seem to have learned not to transfer entire WMD or missile munitions in one shipment, so they are much more likely to transfer components of WMD items under the guise of dual-use items,[20] which populate the AG and MTCR lists.

[16] This regulation was last amended on October 24, 2008. See Council of the European Union, "Council Regulation (EC) No. 1334/2000 of 22 June 2000, Setting up a Community Regime for the Control of Exports of Dual Use Items and Technology (as last amended by Council Regulation (EC) No. 1167/2008, 24 October 2008."

[17] UNSCR 1540 (2004), S/RES/1540 (2004), April 28, 2004; and UNSCR 1673 (2006), S/RES/1673 (2006), April 27, 2006.

[18] This type of provision requires exporters to notify the authorities or obtain a license to export any item, even a noncontrolled item, if they know or are informed that the item will be used by certain countries for prohibited WMD or missile programs.

[19] The Bureau of Industry and Security has issued a list of red flag indicators for export transactions. Key among these indicators are: "the customer or purchasing agent is reluctant to offer information about the end-use of the item"; "the product's capabilities do not fit the buyer's line of business"; "routine installation, training, or maintenance services are declined by the customer"; "the shipping route is abnormal for the product and destination." U.S. Department of Commerce, Bureau of Industry and Security, "Red Flag Indicators: Things to Look for in Export Transactions," Web page, undated.

[20] On May 27, 2008, according to an article in *The Daily Telegraph*, "John Rood, the US Acting Undersecretary of State, said . . . that there had been dozens of PSI interdictions, including preventing the export of dual-use missile-related technologies as well as nuclear-related items to Iran. He gave no details." "Syria-Bound Missile Components Intercepted, Claims US," *The Daily Telegraph*, May 29, 2008.

Once intelligence about an illicit transaction is considered credible, the United States must decide whether shipment of the suspected item as a possible dual-use item is in violation of export controls. If it is, the United States will thwart the illicit trade by first using options that are the least controversial in the international community, with PSI interdiction as the last resort.[21] For cases in which PSI interdiction becomes the only option, the United States will make sure that interdiction is consistent with U.S. domestic laws and international obligations.[22] Even then, it will want to use the least controversial type of PSI interdiction. For example, if U.S. Immigration and Customs Enforcement (ICE) officers notice a suspected shipment of U.S. origin goods is on its way to a foreign country, they can issue a "redelivery order" to the government of that country pursuant to a Customs Mutual Agreement request. Subsequently, the shipment will be redelivered to the United States. This example illustrates that PSI can stop the transfer of WMD items without relying on its most controversial activity, interdiction at high seas.

Let us now assume that the United States decides that an interdiction at sea is necessary. The inspectors will most likely be unable to detect any illicit WMD items at a standoff distance from the suspected ship, and thus will have to board the ship and attempt to locate the suspected WMD items based on the intelligence data they obtained. If they locate the items, they may need to reach-back to the intelligence source in order to identify them. Unfortunately, the suspected items often cannot be found by the boarding party. The National Military Command Center, located in the Joint Staff area of the Pentagon, then makes the decision to detain or release the ship based on its communications with the boarding party.

Because of the myriad difficulties of inspecting individual cargo containers at sea, a thorough inspection has to be conducted in a port. This implies that a suspected ship will have to be escorted to a designated port and that a shipment delay will result. The in-port inspection will be performed with the aid of detectors and inspection procedures. Depending on the types of WMD items suspected, communications with SMEs or even their presence at the inspection port may be necessary. However, in-port searches of ships are generally conducted by host-country personnel rather than U.S. inspectors. These searches may take three to four weeks and require three or four passes, which may make a foreign partner reluctant to search an entire ship because of the disrupting effect on trade. Also, trading countries may not want their ports to acquire a reputation for being difficult to use. Thus, the United States would have to request such inspection judiciously, based on credible and specific intelligence.

Readings for Session 6: Detection of WMD, Their Delivery Systems, and Related Materials

T. R. Twomey and R. M. Keyser, "Hand-Held Radio Isotope Identifiers for Detection and Identification of Illicit Nuclear Materials Trafficking: Pushing the Performance Envelope,"

---

[21]   See the third paragraph in Session 4, International Agreements Relevant to PSI, which discusses the available options, progressing from the least provocative and controversial to the most.

[22]   See Session 4, Internal Agreements Relevant to PSI, Group C: UN Resolutions, for a discussion of how a PSI member can take advantage of UNSCRs—such as UNSCR 1737 against Iran and UNSCR 1718 against North Korea—to justify a PSI interdiction.

undated but probably September 2004. Downloads as of February 21, 2009:
www.ortec-online.com/papers/wco0904.pdf
Hans Binnendijk, Leigh C. Caraher, Timothy Coffey, and H. Scott Wynfield, "The Virtual
Border: Countering Seaborne Container Terrorism," *Defense Horizons*, 16, August 2002. As
of February 21, 2009:
http://www.ndu.edu/inss/DefHor/DH16/DH16.htm

For Further Study

U.S. Government Accountability Office, "Efforts to Deploy Radiation Detection Equipment
in the United States and in Other Countries," testimony statement of Gene Aloise, Director
of Natural Resources and Environment, GAO-05-840T, June 21, 2005.

Victor Orphan, Ernie Muenchau, Jerry Gormley, and Rex Richardson, "Advanced Cargo
Container Scanning Technology Development," Science Applications International Corporation, undated. Downloads as of February 21, 2009:
http://www.trb.org/Conferences/MTS/3A%20Orphan%20Paper.pdf

PSI Exercises and Lessons Learned

The U.S. Department of State fact sheet on the Proliferation Security Initiative (PSI) provides basic information about PSI exercises:[1]

> The PSI provides participating countries with opportunities to improve national capabilities and authorities to conduct interdictions. A robust PSI exercise program allows participants increase their interoperability, improve interdiction decision-making processes, and enhance the interdiction capacities and readiness of all participating states. In five years, PSI partners have sustained one of the only global, interagency, and multinational exercise programs, conducting over 30 operational air, maritime, and ground interdiction exercises involving over 70 nations. These exercises are hosted throughout the world by individual PSI participants and consist of air, maritime, and ground exercises executed by participants' interagency and ministries focusing on improving coordination mechanisms to support interdiction-related decision-making.
>
> Furthermore, the PSI Operational Experts Group (OEG), a group of military, law enforcement, intelligence, legal, and diplomatic experts from twenty PSI participating states, meets regularly to develop operational concepts, organize the interdiction exercise program, share information about national legal authorities, and pursue cooperation with key industry sectors. The OEG works on behalf of *all* PSI partners and works enthusiastically to share its insights and experiences through bilateral and multilateral outreach efforts.

The 20 members of the OEG are Argentina, Australia, Canada, Denmark, France, Germany, Greece, Italy, Japan, the Netherlands, New Zealand, Norway, Poland, Portugal, Russia, Singapore, Spain, Turkey, the United Kingdom, and the United States.

The details of many PSI exercises are not publicly available, but a fairly detailed report is available for "Pacific Shield 07," a maritime interdiction exercise hosted by Japan.[2] The countries providing assets to the exercise were Australia, France, Japan, New Zealand, Singapore, the United Kingdom, and the United States. In addition, 34 countries took part as observers:

- Asia and Oceania: Brunei, India, Laos, Malaysia, Marshall Islands, Mongolia, Pakistan, Papua New Guinea, Philippines, and Vietnam

---

[1]   U.S. Department of State, Under Secretary for Arms Control and International Security, Bureau of International Security and Nonproliferation, *Proliferation Security Initiative (PSI)*, Fact Sheet, Bureau of International Security and Nonproliferation, Washington, D.C., May 26, 2008.

[2]   Ministry of Foreign Affairs of Japan, *PSI Maritime Interdiction Exercise "Pacific Shield 07" Hosted by the Government of Japan (Overview and Evaluation)*, October 18, 2007.

- North America: Canada
- Central and South America: Brazil, Chile, El Salvador, Panama
- Middle East: Bahrain, Israel, Jordan, Oman, Qatar, Turkey, United Arab Emirates
- Europe and other regions: Denmark, Germany, Italy, Kazakhstan, the Netherlands, Norway, Poland, Romania, Russia, Slovenia, Spain, Ukraine.

The exercise took place over three days. Events on the first day, October 13, 2007, took place at sea east of Oshima Island. Each of the seven countries that had provided assets to the exercise took turns demonstrating the sequence of search, detection, and tracking followed by boarding.

Events on the second day, October 14, 2007, took place in Yokosuka Shinko Port. Each of six countries (Australia, France, Japan, Singapore, the United Kingdom and the United States) took turns conducting a boarding inspection. The elements of this inspection included boarding, onboard search, and the detection of suspect material. For the purposes of the exercise, the target vessel was moored to a pier in the port but assumed to be at sea.

Events on the last day, October 15, 2007, took place in Yokohama Port. Four countries (Australia, Japan, Singapore, and the United States) took turns demonstrating a sequence of inspection measures at the port. The specific scenario for this day was that the Japanese government had learned that sodium cyanide, which can be used for the production of a nerve agent, was going to be transshipped at Yokohama Port and placed on a vessel destined for a country of proliferation concern. In the exercise, the vessel was inspected and the material was detected and seized.

In its report on the three-day exercise, Japan published the following overall evaluation[3]

(1) Expression of Strong Will and Determination of Counter-Proliferation

Holding the second* PSI Exercise in Japan demonstrated the will and determination of Japan as well as the international community to tackle the proliferation of weapons of mass destruction (WMDs) and related materials. The whole Exercise program was open to observers and media, and was useful for promoting the understanding about the goals and activities of the PSI.

—

*Japan hosted its first PSI Maritime Interdiction Exercise "Team Samurai 04" in October 2004.

(2) Variety of Activities by Participants, Strengthening Coordination between Related Agencies

Pacific Shield 07 adopted "capability demonstration" approach, in which all asset dispatching countries were provided with sufficient time and opportunities for practical exercise and opportunities to witness what other countries would do. This approach was useful for mutual understanding and exchange between related agencies from participating/observer countries, and contributed to further promotion of the effectiveness of counter-proliferation measures.

---

[3]   Ministry of Foreign Affairs of Japan, 2007.

In addition to the search, detection, tracking and boarding inspection exercises, which also took place in Team Samurai 04, the port inspection exercise was conducted for the first time in the exercise hosted by Japan. As for Japan's law-enforcement exercise, the coordination between the Police, the Customs and the Coast Guard was strengthened through the Exercise. Also, 101 NBC Protection Unit of the JGSDF conducted exercise regarding decontamination.

(3) Active Participation by a Wider Range of Countries

New Zealand, Singapore and the United Kingdom participated for the first time in Japan-hosted PSI exercise with their assets. Also, observers from a wide range of countries, including countries in Asia-Oceania and Middle-East, as well as non PSI supporting countries, participated in Pacific Shield 07. The number of countries increased significantly compared to Team Samurai 04*.

Observers from those countries had active exchanges with other participants/observers during the 3-day Exercise and related events, and deepened their understanding about the PSI, the importance of counter-proliferation efforts and the related policies of other countries.
—
*Team Samurai 04 were participated by 21 countries, including 3 asset dispatching countries (Australia, France and the US).

The U.S. Department of State's PSI fact sheet states that "over 30 operational air, maritime, and ground interdiction exercises" have been conducted as of May 26, 2008.[4] The official State department calendar of events Web page indicates that 36 interdiction exercises have taken place through 2008.[5] It also lists, for each event, the host country, location, and type of interdiction (see table, below), along with upcoming events, four of which are exercises planned for 2009.

Readings for Session 7: PSI Exercises and Lessons Learned

Ministry of Foreign Affairs of Japan, *PSI Maritime Interdiction Exercise "Pacific Shield 07" Hosted by the Government of Japan (Overview and Evaluation)*, October 18, 2007. As of February 21, 2009:
http://www.mofa.go.jp/policy/UN/disarmament/arms/psi/overview0710.html

[4] U.S. Department of State, Under Secretary for Arms Control and International Security, Bureau of International Security and Nonproliferation, *Proliferation Security Initiative (PSI)*, 2008.

[5] U.S. Department of State, Under Secretary for Arms Control and International Security, Bureau of International Security and Nonproliferation, "Calendar of Events: Proliferation Security Initiative (PSI) Exercises," Web page, current as of September 2008.

PSI Exercises Conducted Through December 2008

Name	Date	Type	Host	Location
PACIFIC PROTECTOR	09/03	Maritime interdiction	Australia	Coral Sea
Air CPX	10/03	Tabletop exercise	UK	Mediterranean
SANSO '03	10/03	Maritime interdiction	Spain	Mediterranean
BASILIC '03	11/03	Maritime interdiction	France	Mediterranean
SEA SABER	01/04	Maritime interdiction	U.S.	Arabian Sea
AIR BRAKE '03	02/04	Maritime interdiction	Italy	Mediterranean
HAWKEYE	04/04	Ground	Germany	Germany
SAFE BORDERS	04/04	Ground	Poland	Poland
CLEVER SENTINEL	04/04	Maritime interdiction	Italy	Mediterranean
APSE '04	06/04	Air Interception	France	France
	09/04	Maritime interdiction game	U.S.	U.S.
TEAM SAMURAI '04	10/04	Maritime interdiction	Japan	Japan
CHOKEPOINT '04	11/04	Maritime interdiction	U.S.	U.S.
NINFA '05	04/05	Maritime/ground interdiction	Portugal	Portugal
BOHEMIAN GUARD '05	06/05	Ground interdiction	Czech Republic, Poland	Czech Republic
BLUE ACTION '05	06/05	Air/ground interdiction	Spain	Mediterranean
DEEP SABRE	08/05	Maritime/ground interdiction	Singapore	Singapore
	10/05	Air interdiction game	Norway	Norway
EXPLORING THEMIS	11/05	Maritime/ground interdiction	UK	Indian Ocean
TOP PORT	04/06	Maritime interdiction	Netherlands	Netherlands
PACIFIC PROTECTOR '06	04/06	Air/ground interdiction	Australia	Australia
ANATOLIAN SUN	05/06	Air/ground/maritime	Turkey	Turkey
HADES '06	06/06	Air interdiction	France	France
AMBER SUNRISE	09/06	Maritime/ground interdiction	Poland	Poland
LEADING EDGE	10/06	Maritime/ground interdiction	U.S.	Persian Gulf
SMART RAVEN	04/07	Air interdiction	Lithuania	Lithuania
ADRIATIC GATE	05/07	Ground/port interdiction	Slovenia	Slovenia
	06/07	Maritime interdiction game	U.S.	U.S.
PANAMAX '07	08/07	Maritime interdiction	U.S.	Panama
PACIFIC SHIELD '07	10/07	Maritime/port interdiction	Japan	Japan
EASTERN SHIELD '07	10/07	Air/ground/maritime interdiction	Ukraine	Ukraine
GUISTIR '08	03/08	Maritime/port interdiction	Djibouti, France	Djibouti
PHOENIX EXPRESS '08	04/08	Maritime interdiction	U.S.	Mediterranean
ADRIATIC SHIELD '08	05/08	Maritime interdiction	Croatia	Croatia
PANAMAX '08	08/08	Maritime interdiction	U.S.	*Not available*
MARU	09/08	*Not available*	New Zealand	*Not available*

SOURCES: (1) U.S. Department of State, Bureau of International Security and Nonproliferation, Under Secretary for Arms Control and International Security, Bureau of International Security and Nonproliferation, "Calendar of Events: Proliferation Security Initiative (PSI) Exercises," Web page, current as of September 2008. (2) "PSI Exercises," Web page, undated, provides links to sites with full descriptions of some of the exercises. (3) Ministry of Foreign Affairs of Republic of Poland, "Krakow Initiative: Proliferation Security Initiative," Web page, © 2005, lists PSI exercises in Central and Eastern Europe and provides links to further information on the exercises. (4) Ministry of Foreign Affairs of Japan, *PSI Maritime interdiction Exercise "Pacific Shield 07" Hosted by the Government of Japan (Overview and Evaluation)*, October 18, 2007.

Responding to Issues Challenging PSI

This session addresses the issues that have affected some countries' disposition toward endorsing the Proliferation Security Initiative (PSI). We discuss these challenges to PSI, which various countries have voiced or implied at various time, as well as questions about PSI that appear in the assigned additional readings for this manual.

In approaching these issues, it is worth noting that none of the countries that have withheld affiliation with PSI opposes the nonproliferation objectives of PSI. For example, China's foreign ministry has stated that

> China is firmly opposed to proliferation of WMD and their means of delivery and stands for the attainment of the non-proliferation goal through political and diplomatic means. We understand the concern of PSI participants over the proliferation of WMD and their means of delivery, and share the non-proliferation goal of the PSI.[1]

However, the five holdout countries—that is, the five key countries that have so far chosen not to affiliate with PSI: Indonesia, Malaysia, India, Pakistan, and China—and others have raised numerous arguments that have led to an unwillingness to affiliate with PSI and uneasiness about PSI's interdiction practices.[2] These arguments, which we discuss in the following sections, along with appropriate responses, can be grouped under three headings:

1. the law of the sea, and the right of innocent passage
2. ambiguity about PSI interdiction circumstances
3. U.S. dominance of PSI and related notion that affiliation with PSI implies a broader than desired acceptance of and association with U.S. foreign and defense policies.

Law of the Sea and the Right of Innocent Passage

China's main issue related to law of the sea and the right of innocent passage has been the legitimacy and consequences of interdiction. Its argument is that PSI could infringe upon the right of innocent passage of Chinese or other state-flagged ships through the territorial waters of PSI members, such as the Straits of Malacca. Thus, China is concerned that PSI members might

---

[1] Ministry of Foreign Affairs of the People's Republic of China, *Proliferation Security Initiative*, May 21, 2007. Also, see Charles Wolf, Jr., Brian G. Chow, and Gregory S. Jones, *Enhancement by Enlargement: The Proliferation Security Initiative*, MG-806-OSD, Santa Monica, Calif.: RAND, 2008, pp. 27–28.

[2] For a full discussion of the holdout countries and the related issues, see Wolf, Chow, and Jones, 2008, Chapter Two.

be subject to interdiction in situations that China itself considers to be "innocent passage," which would constitute a violation of international law. Indonesia, another holdout country, has expressed the same concern.

This argument also encompasses a concern that PSI commitments would violate the stipulation in the UN Convention on the Law of the Sea (UNCLOS) that protects the right of innocent passage—namely, the right of appropriately state-flagged vessels to transit international waters without question or interruption.[3]

In forming a response to this argument, it is important to note that UN Security Council Resolution (UNSCR) 1540, adopted in 2004, already obligates all states to "adopt and enforce appropriate effective laws which prohibit any non-state actor to manufacture, acquire, possess, develop, transport, transfer, or use nuclear, chemical, or biological weapons and their means of delivery."[4] It can be argued that UNSCR 1540 itself raises a possible inconsistency between it and the law of the sea right of passage. For example, if, on the basis of credible intelligence, a state-flagged vessel were suspected of transporting WMD that might be or might be suspected of being linked to a non-state actor, or of not being able to prevent a non-state actor from acquiring the transported cargo once off-loaded, UNSCR 1540 would already provide grounds for suspending the specified law of the sea right of innocent passage.

In other words, the right of innocent passage is not unqualified. Indeed, UNCLOS (Article 19) explicitly introduces general qualifications to the right of innocent passage by plainly acknowledging that passage is not "innocent" if it is "prejudicial to the peace, good order, or security of the coastal state" or is not "in conformity . . . with other rules of international law."

Hence, it may be justifiable to interpret the illegitimate transport of WMD or missile items as prejudicial to peace or security, thereby voiding the right of innocent passage.

Ambiguity About PSI Interdiction Circumstances

This second argument against PSI affiliation is based on concern that the circumstances in which interdiction would be applied are ambiguous. For example, one of the PSI interdiction principles states:

> Of their own initiative, or at the request and good cause shown by another state, [PSI participants will] take action to board and search any vessel flying their flag in their internal waters or territorial seas, or areas beyond the territorial seas of any other state, that is reasonably suspected of transporting such [WMD-related] cargoes to or from states or non-state actors of proliferation concern, and to seize such cargoes that are identified."[5]

In this case, the argument arises from the seeming breadth of this principle and from some countries' acute concern that invoking and applying it might violate international law.

---

[3]   *United Nations Convention on the Law of the Sea of 10 December 1982.*

[4]   UNSCR 1540 (2004), S/RES/1540 (2004), April 28, 2004.

[5]   U.S. Department of State, Under Secretary for Arms Control and International Security, Bureau of International Security and Nonproliferation, *Interdiction Principles for the Proliferation Security Initiative*, Bureau of International Security and Nonproliferation, Washington, D.C., September 4, 2003.

This ambiguity-based argument against PSI affiliation can be readily countered by noting that PSI is not a formal treaty-based organization and hence does not obligate PSI affiliates to take specific actions at certain times.[6] While PSI provides a basis for cooperation among partners on specific actions when the need to deter and stop proliferation arises, none of the PSI interdiction principles obligates affiliates to take any specific action. Any decision by a state to participate in an interdiction effort is completely voluntary.

U.S. Dominance of PSI and Related Implications of Affiliation

The argument against PSI affiliation because of its dominance by the United States sometimes includes the concern that affiliation will imply or be perceived as implying a broader acceptance of U.S. foreign and defense policies than some countries may wish. This argument is also associated with concern that PSI affiliation could be considered an antagonistic action by the primary "states of proliferation concern" (namely, Iran and North Korea) and hence would vitiate economic and political interests and relations that a potential PSI participant has or expects to have with those two states.

The best counters for this argument are the remarkable evolution and expansion of PSI. Since 2003, when PSI was announced by U.S. President George Bush in Poland and its interdiction principles were formulated at meetings held in Spain, Australia, and France, the initiative has expanded to include 93 countries. The partnership and cooperation of these countries are led by a group of 20 rotating members of the PSI Operational Experts Group (OEG), many of whom frequently and substantially disagree with U.S. policies. The United States is, to be sure, a major participant in PSI, but it no longer dominates PSI's activities or its policies.

In consequence, the distinctly multilateral character of the initiative's evolution is a strong indication that U.S. dominance is a thing of the past. And PSI's present and future emphatically represent multilateral cooperation among 93 partners.

Readings for Session 8: Responding to Issues Challenging PSI

Michael Byers, "Policing the High Seas: The Proliferation Security Initiative," *The American Journal of International Law*, 98(3), July 2004, pp. 526–544.

John Duff, "A Note on the United States and the Law of the Sea: Looking Back and Moving Forward," *Ocean Development and International Law*, 15, 2004, pp. 195–219.

Mark J. Valencia, *The Proliferation Security Initiative: Making Waves in Asia*, Adelphi Paper 376, International Strategic Studies Institute, 2005. Downloadable as of February 22, 2009, at:
http://www.iiss.org/publications/adelphi-papers/about-adelphi-papers/

---

[6]   U.S. Department of State, Under Secretary for Arms Control and International Security, Bureau of International Security and Nonproliferation, *Interdiction Principles for the Proliferation Security Initiative*, 2003.

Charles Wolf, Jr., Brian G. Chow, and Gregory S. Jones, *Enhancement by Enlargement: The Proliferation Security Initiative*, MG-806-OSD, Santa Monica, Calif.: RAND, 2008. As of February 19, 2009:
http://www.rand.org/pubs/monographs/MG806/

Andrew C. Winner, "The Proliferation Security Initiative: The New Face of Interdiction," *Washington Quarterly*, Spring, 2005. Downloads as of February 19, 2009:
http://www.twq.com/05spring/docs/05spring_winner.pdf

Benjamin Friedman, "The Proliferation Security Initiative: The Legal Challenge," policy brief written for Bipartisan Security Group, a program of Global Security Group, Washington, D.C., September 4, 2003. Downloads as of February 22, 2009:
www.gsinstitute.org/gsi/pubs/09_03_psi_brief.pdf

C. Raja Mohan, "Dismantling Core Group: US Eases India's Path to Proliferation Security," *Indian Express*, New Delhi, August 18, 2005.

Enhancing Capabilities for PSI Participation

The Proliferation Security Initiative (PSI) is a multilateral endeavor whose purpose is to prevent or at least inhibit the spread of weapons of mass destruction (WMD), their delivery systems, and related materials (which we refer to collectively as *WMD items*) to and from states or non-state actors whose possession of them would be a serious threat to global or regional security. PSI activities are led by PSI's Operational Experts Group (OEG), which comprises 20 countries that plan and implement the exercises and other multilateral efforts designed to further PSI's purpose.

Affiliation with PSI is entirely voluntary. It is neither mandated nor precluded by a country's membership in other multilateral or bilateral organizations—for example, the UN, the North Atlantic Treaty Organization (NATO), and the Shanghai Cooperation Organization (SCO).

Enhancement of Participants' Capabilities

Once a state affiliates with PSI, the extent, frequency, and intensity of its participation in PSI's cooperative and partnering activities are also voluntary. However, even though the extent of participation is voluntary, the capabilities that participants bring to the table for the PSI exercises and discussions can be honed and enhanced, thereby enabling their contributions to be correspondingly improved. Specifically, PSI provides participating countries with opportunities to increase and modernize their customs and invoicing practices, to improve their technical inspection and detection capabilities, to expedite the rapid exchange of relevant information concerning suspected proliferation activity, and to realize these benefits while protecting the confidential character of classified information provided by other participating states.

Furthermore, through its robust exercise programs, PSI allows participants to increase their interoperability and improve their interdiction decisionmaking processes, and enhances the capacities and readiness of all participating states for identifying and, where necessary, interdicting transshipments of WMD items. Over five years, PSI partners have sustained one of the only global, interagency, and multinational exercise programs, conducting over 30 operational air, maritime, and ground interdiction exercises involving more than 70 nations. These exercises have been hosted throughout the world by individual PSI participants and have been executed by those participants' agencies and ministries in cooperation with the corresponding entities of other participating countries. The focus in these cooperative efforts is on improving coordination mechanisms to support decisionmaking consonant with PSI's purposes.

Enhancing PSI's Effectiveness: Other Lines of Inquiry and Research

In confronting the possibility of a more-proliferated world, several additional lines of inquiry are worth consideration and discussion. Some of these are briefly discussed in the following sections.

Cooperation with Private Industry

Some of the most effective PSI strategies have involved engaging the financial industry to stop shipments of WMD items. By freezing financial assets and transactions known to facilitate the proliferation of WMD items, the U.S. government and the international community have worked together to complicate and stop this dangerous trade. It may also be useful to discuss with the insurance industry whether and, if so, how premiums charged for insuring cargo (whether transported by surface, air, or sea) take into account PSI affiliation of the transport vehicle's nation of origin. The premise underlying this inquiry is that processes learned and experience gained through participation in PSI and its exercises may reduce various risks, such as those associated with possible accidents and delays connected with the transport of WMD items.

Consequently, the risk exposure of PSI participants might be calculably less than the risk exposure of nonparticipants. In effect, the transport of WMD systems or components involves increased risks to the points of origin, destination, and thoroughfare. To the extent this is true, PSI affiliation may carry with it some presumptive reduction of these risks and hence a warrantable reduction in the insurance premium charged to the flagged ships or transport media of PSI participants.

An Interdiction Compensation Fund?

The right of innocent passage is an issue that has frequently arisen in connection with PSI.[1] Part of the concern about this issue is that even an innocent ship might suffer delay by interdiction. To allay this concern, PSI might consider establishing a fund to compensate the owner(s) of a carrier (whether ship, ground vehicle, or airplane) and the cargo recipient(s) affected by the delay, if a particular interdiction turned out to be based on erroneous intelligence and the carrier was not at fault. Any compensation awarded would be intended to cover basic losses resulting from the delay in cargo delivery and to avoid any windfall resulting from such a mistake.

An issue for further consideration is whether it would be useful to establish such a fund and, if so, how it might be structured and financed and how it would operate. Another issue is how to establish the legal authorities so that the potential compensation would preclude frivolous lawsuits and claims. Consideration of these issues should be pursued multilaterally within PSI, perhaps under the auspices of the OEG.

Differing Interpretations of the Right of Innocent Passage

One of the obstacles to enlarging PSI is concern on the part of some countries that their interpretation of their right of innocent passage under the UN Convention on the Law of the Sea (UNCLOS) may differ from that of other countries. Specifically, these countries may be concerned that if they were a PSI member, their ships would be interdicted by other PSI members. For example, flagged nation B may have a plausible concern that littoral state A might interpret

---

[1]   This topic is discussed in more detail in Sessions 4, 5, and 8.

its obligations to interdict B's ship without B's consent as long as the interdiction is conducted in A's territorial waters and is preceded by reasonable grounds for suspicion. In practice, however, a PSI participant will not interdict other PSI participants' ships in its territorial seas, or on the high seas, without those participants' permission. PSI's OEG should make this point explicit, thereby helping to resolve what is viewed as a key obstacle by several nations that might otherwise affiliate with PSI.

Detection Technology

It would be useful to conduct a survey of the status and trends in WMD detection and sensing technologies, since improvements in these may enable more accurate and more rapid identification of WMD items, thereby enhancing PSI's future effectiveness. In addition, increasing PSI participants' access to advancements in sensing and detection technologies would increase the attractiveness of PSI participation for both current and prospective affiliates. It would also increase the contributions of the affiliates' cooperative efforts to the effectiveness and success of PSI activities.

References

Ahlstrom, Christer, "The Proliferation Security Initiative: International Law Aspects of the Statement of Interdiction Principles," Chapter 18, S*IPRI Yearbook 2005: Armaments, Disarmament and International Security*, United Kingdom: Oxford University Press, 2005. Downloadable as of February 19, 2009, at: http://yearbook2005.sipri.org/ch18/ch18

[The] Atomic Energy Act of 1954 (as amended), Public Law 83-703, 68 Stat. 919, August 30, 1954, Title I: Atomic Energy. Downloadable as of February 21, 2009, at: http://www.nrc.gov/about-nrc/governing-laws.html

Australia Group, "Australia Group Common Control Lists," © 2007. As of February 18, 2009: http://www.australiagroup.net/en/controllists.html

Binnendijk, Hans, Leigh C. Caraher, Timothy Coffey, and H. Scott Wynfield, "The Virtual Border: Countering Seaborne Container Terrorism," *Defense Horizons*, 16, August 2002. As of February 21, 2009: http://www.ndu.edu/inss/DefHor/DH16/DH16.htm

Boese, Wade, "Key U.S. Interdiction Initiative Claim Misrepresented," *Arms Control Today*, July/August 2005, pp. 26–27. Full article (unpaged) as of February 19, 2009: http://www.armscontrol.org/act/2005_07-08/Interdiction_Misrepresented

Bush, George W., speech given in Krakow, Poland, May 31, 2003. Full transcript available in "Bush Urges NATO Nations to Unite in Fight Against Terrorism," May 31, 2003. As of February 28, 2009: http://www.globalsecurity.org/military/library/news/2003/05/mil-030531-usia03.htm

Byers, Michael, "Policing the High Seas: The Proliferation Security Initiative," *The American Journal of International Law*, 98(3), July 2004, pp. 526–544.

Center for Nonproliferation Studies, "Inventory of International Nonproliferation Organizations and Regimes," undated. As of February 20, 2009: http://cns.miis.edu/inventory/index.htm

Chairman of the Joint Chiefs of Staff, *Proliferation Security Initiative (PSI) Activity Program*, CJCS Instruction 3520.02A, March 1, 2007 (current as of May 20, 2008). Downloads as of February 19, 2009: http://www.dtic.mil/cjcs_directives/cdata/unlimit/3520_02.pdf

Council of the European Union, "Council Regulation (EC) No. 1334/2000 of 22 June 2000, Setting up a Community Regime for the Control of Exports of Dual Use Items and Technology (as last amended by Council Regulation (EC) No. 1167/2008, 24 October 2008." As of February 27, 2009: http://www.consilium.europa.eu/showPage.aspx?id=408&lang=en

"Customs Seek Help of Experts," *Indian Express*, July 1, 1999. As of February 19, 2009: http://www.expressindia.com/news/ie/daily/19990702/ige02125.html

Duff, John, "A Note on the United States and the Law of the Sea: Looking Back and Moving Forward," *Ocean Development and International Law*, 15, 2004, pp. 195–219.

ElBaradei, Mohamed, "7 Steps for Preventing Nuclear Proliferation," February 15, 2005. As of February 21, 2009: http://www.asahi.com/english/opinion/TKY200502150114.html

Friedman, Benjamin, "The Proliferation Security Initiative: The Legal Challenge," policy brief written for Bipartisan Security Group, a program of Global Security Group, Washington, D.C., September 4, 2003. Downloads as of February 22, 2009:
www.gsinstitute.org/gsi/pubs/09_03_psi_brief.pdf

Hawkins, William, "Chinese Realpolitik and the Proliferation Security Initiative," February 18, 2005. As of February 11, 2009:
http://www.asianresearch.org/articles/2505.html

International Civil Aviation Organization, "Covention on International Civil Aviation," Doc 7300/9 (9th edition), 2006. Downloads as of February 19, 2009:
http://www.icao.int/icaonet/dcs/7300_cons.pdf

Jofi, Josif, "The Proliferation Security Initiative: Can Interdiction Stop Proliferation?" *Arms Control Today*, June 2004. As of February 21, 2009:
http://www.armscontrol.org/act/2004_06/Joseph

Kerr, Paul, "Libya Vows to Dismantle WMD Program," *Arms Control Today*, January/February 2004, p. 29. Full article (unpaged) as of February 19, 2009:
http://www.armscontrol.org/act/2004_01-02/Libya

Knowlton, Brian, "Ship Allowed to Take North Korea Scuds on to Yemeni Port: U.S. Frees Freighter Carrying Missiles," *International Herald Tribune*, December 12, 2002. As of February 19, 2009:
http://www.iht.com/articles/2002/12/12/scuds_ed3_.php

Logan, Samuel E., "The Proliferation Security Initiative: Navigating the Legal Challenges," *Journal of Transnational Law & Policy*, 14(2), Spring 2005. Downloadable as of February 28, 2009, at:
http://www.law.fsu.edu/journals/transnational/backissues/issue14_2.html

Ministry of Foreign Affairs of Japan, *PSI Maritime Interdiction Exercise "Pacific Shield 07" Hosted by the Government of Japan (Overview and Evaluation)*, October 18, 2007. As of February 21, 2009:
http://www.mofa.go.jp/policy/UN/disarmament/arms/psi/overview0710.html

Ministry of Foreign Affairs of Republic of Poland, "Krakow Initiative: Proliferation Security Initiative," Web page, © 2005, lists PSI exercises in Central and Eastern Europe and provides links to further information on the exercises. As of February 22, 2009:
http://www.psi.msz.gov.pl/index.php?document=54

Ministry of Foreign Affairs of the People's Republic of China, *Proliferation Security Initiative*, May 21, 2007. As of February 28, 2009:
http://www.fmprc.gov.cn/eng/wjb/zzjg/jks/kjlc/fkswt/dbfks/t321019.htm

———, "Statement by the Ministry of Foreign Affairs of the People's Republic of China on the 'Yin He' Incident, dated 4 September 1993." As of February 19, 2009:
http://www.nti.org/db/china/engdocs/ynhe0993.htm

Missile Technology Control Regime, "Objectives of the MTCR," undated. As of February 18, 2009:
http://www.mtcr.info/english/objectives.html

———, *Equipment, Software and Technology Annex*, updated November 5, 2008. Downloadable as of February 18, 2009, at:
http://www.mtcr.info/english/annex.html

Mohan, C. Raja, "Dismantling Core Group: US Eases India's Path to Proliferation Security," *Indian Express*, New Delhi, August 18, 2005.

National Nuclear Security Administration, "Megaports Initiative," undated. As of February 29, 2009:
http://nnsa.energy.gov/nuclear_nonproliferation/1641.htm

National reports on the implementation of Security Council Resolution 1540, submitted to the Chairman of the Security Council Committee established pursuant to Resolution 1540 (2004). (Note: These are equivalent to "United States Report to the Committee Established Pursuant to Resolution 1540 (2004): Efforts Regarding Security Council Resolution 1540 (2004)," S/AC.44/2004/(02)/5, shown, below, in this list.) Downloadable by country as of February 13, 2009, at:
http://www.un.org/sc/1540/nationalreports.shtml

National Strategy to Combat Weapons of Mass Destruction, NSPD 17, December 2002. As of February 19, 2009:
http://www.globalsecurity.org/security/library/policy/national/hspd-4.htm

Nuclear Threat Initiative, *A Tutorial on Nuclear Weapons and Nuclear-Explosive Materials—Part Five*, 2005.

Orphan, Victor, Ernie Muenchau, Jerry Gormley, and Rex Richardson, "Advanced Cargo Container Scanning Technology Development," Science Applications International Corporation, undated. Downloads as of February 21, 2009:
http://www.trb.org/Conferences/MTS/3A%20Orphan%20Paper.pdf

Proliferation Security Initiative: Statement of Interdiction Principles, Paris, September 4, 2003. As of February 25, 2009:
http://www.proliferationsecurity.info/principles.html

"PSI Exercises," Web page, undated, provides links to sites with full descriptions of the exercises. As of February 22, 2009:
http://www.proliferationsecurity.info/exercises.html

Ronzitti, Natalie, "The Law of the Sea and the Use of Force Against Terrorist Activities," in Natalie Ronzitti (ed.), *Maritime Terrorism and International Law*, Netherlands: Kluwer Law International, 1990, pp. 1–15.

Shanker, Thom, "US Remains Leader in Global Arms Sales, Report Says," *New York Times*, September 25, 2003, p. A-12.

Song, Yann-hui, "An Overview of Regional Responses in the Asia-Pacific to the PSI," in *Countering the Spread of Weapons of Mass Destruction: The Role of the Proliferation Security Initiative*", Pacific Forum CSIS's *Issues & Insights*, 4(5), July 2004, pp. 7–31.

Squassoni, Sharon, "Proliferation Security Initiative (PSI)," *CRS Report for Congress*, RS21881, September 14, 2006. As of February 25, 2009:
http://opencrs.com/document/RS21881/2005-06-07

Stockholm International Peace Research Institute, *SIPRI Yearbook 2005: Armaments, Disarmament and International Security*, United Kingdom: Oxford University Press, 2005, pp. 640 and 748.

Suettinger, Robert L., *Beyond Tiananmen: The Politics of U.S.-China Relations, 1989–2000*, Brookings Institution Press, Washington D.C., 2003, pp. 174–177.

"Syria-Bound Missile Components Intercepted, Claims US," *The Daily Telegraph*, May 29, 2008. As of February 22, 2009:
http://www.telegraph.co.uk/news/worldnews/2045825/Syria-bound-missile-components-intercepted-claims-US.html

Twomey, T. R., and R. M. Keyser, "Hand-Held Radio Isotope Identifiers for Detection and Identification of Illicit Nuclear Materials Trafficking: Pushing the Performance Envelope," undated but probably September 2004. Downloads as of February 21, 2009:
www.ortec-online.com/papers/wco0904.pdf

United Nations Convention on the Law of the Sea of 10 December 1982. Downloadable as of February 19, 2009, at:
http://www.un.org/Depts/los/convention_agreements/convention_overview_convention.htm

United Nations, "International Convention on the Suppression of Acts of Nuclear Terrorism," 2005. Downloads as of February 25, 2009:
http://untreaty.un.org/English/Terrorism/English_18_15.pdf

UNSCR 1540 (2004), S/RES/1540 (2004), April 28, 2004. Downloadable as of February 19, 2009, at:
http://www.un.org/sc/1540/

UNSCR 1673 (2006), S/RES/1673 (2006), April 27, 2006. Downloadable as of February 19, 2009, at:
http://www.un.org/sc/1540/

UNSCR 1718 (2006), S/RES/1718 (2006), October 14, 2006. Downloadable as of February 20, 2009, at:
http://www.un.org/sc/committees/1718/resolutions.shtml

UNSCR 1737 (2006), S/RES/1737 (2006), December 27, 2006 (reissue of December 23 version). Downloadable as of February 20, 2009, at:
http://www.un.org/sc/committees/1737/resolutions.shtml

UNSCR 1803 (2008), S/RES/1803 (2008), March 3, 2008. Downloadable as of February 20, 2009, at:
http://www.un.org/sc/committees/1737/resolutions.shtml

U.S. Department of Commerce, Bureau of Industry and Security, "Export Administration Regulations," Web site, last updated January 16, 2009. As of February 19, 2009:
http://www.access.gpo.gov/bis/index.html

———, "Red Flag Indicators: Things to Look for in Export Transactions," Web page, undated. As of February 22, 2009:
http://www.bis.doc.gov/Enforcement/redflags.htm

———, "The Commerce Control List," Part 774 of the Export Administration Regulations (EAR). Available to subscribers to EAR database; various portions downloadable as of February 18, 2009, at:
http://www.access.gpo.gov/bis/ear/ear_data.html

U.S. Department of Homeland Security, "Container Security Initiative Ports," Web page, last reviewed/ modified October 20, 2008. As of February 21, 2009:
http://www.dhs.gov/xprevprot/programs/gc_1165872287564.shtm

U.S. Department of State, Under Secretary for Arms Control and International Security, Bureau of International Security and Nonproliferation, "Calendar of Events: Proliferation Security Initiative (PSI) Exercises," Web page, current as of September 2008. As of February 25, 2009:
http://www.state.gov/t/isn/c27700.htm

———, *Interdiction Principles for the Proliferation Security Initiative*, Bureau of International Security and Nonproliferation, Washington, D.C., September 4, 2003. (Note: For the full statement of the principles, which includes introductory text along with the principles, see *Proliferation Security Initiative: Statement of Interdiction Principles*, September 4, 2003, above.) As of February 25, 2009:
http://www.state.gov/t/isn/c27726.htm

———, *Proliferation Security Initiative (PSI)*, Fact Sheet, Bureau of International Security and Nonproliferation, Washington, D.C., May 26, 2008. As of February 25, 2009:
http://www.state.gov/t/isn/115494.htm

———, *Proliferation Security Initiative Frequently Asked Questions (FAQ)*, Fact Sheet, May 22, 2008. As of February 25, 2009:
http://www.state.gov/t/isn/115491.htm

———, "Proliferation Security Initiative Participants," Web page, current as of January 22, 2009. As of February 25, 2009:
http://www.state.gov/t/isn/c27732.htm

———, "Ship Boarding Agreements," Web page, with links to U.S.-country agreements (e.g., "Proliferation Security Initiative Ship Boarding Agreement with Belize"), undated. As of February 25, 2009:
http://www.state.gov/t/isn/c27733.htm

U.S. Department of the Treasury, "Entity List," Supplement No. 4 to Part 744 of the Export Administration Regulations, December 5, 2008. As of February 18, 2009:
http://www.bis.doc.gov/entities/default.htm

U.S. Government Accountability Office, "Efforts to Deploy Radiation Detection Equipment in the United States and in Other Countries," testimony statement of Gene Aloise, Director of Natural Resources and Environment; GAO-05-840T, June 21, 2005.

U.S. Immigration and Customs Enforcement, "About: Offices Within ICE," last modified December 8, 2008. As of February 19, 2009:
http://www.ice.gov/about/operations.htm

"United States Report to the Committee Established Pursuant to Resolution 1540 (2004): Efforts Regarding Security Council Resolution 1540 (2004)," S/AC.44/2004/(02)/5, annex to letter dated October 12, 2004, to Chairman of UN Security Council Committee, October 14, 2004. Downloadable as of February 19, 2009, at: http://www.un.org/Docs/journal/asp/ws.asp?m=S/AC.44/2004/(02)/5

Valencia, Mark J., "The Proliferation Security Initiative: A Glass Half-Full," *Arms Control Today*, June 2007. As of February 19, 2009: http://www.armscontrol.org/act/2007_06/Valencia

———, *The Proliferation Security Initiative: Making Waves in Asia*, Adelphi Paper 376, International Strategic Studies Institute, 2005. Downloadable as of February 22, 2009, at: http://www.iiss.org/publications/adelphi-papers/about-adelphi-papers/

Warrick, Joby, "On North Korean Freighter, a Hidden Missile Factory," *Washington Post*, August 14, 2003. As of February 19, 2009: http://www.washingtonpost.com/ac2/wp-dyn/A56111-2003Aug13?language=printer

Winner, Andrew C., "The Proliferation Security Initiative: The New Face of Interdiction," *Washington Quarterly*, Spring 2005. Downloads as of February 19, 2009: http://www.twq.com/05spring/docs/05spring_winner.pdf

Wolf, Charles, Jr., *Markets or Governments: Choosing Between Imperfect Alternatives*, Chapter 2, "Market Failure," N-2505-SF, Santa Monica, Calif.: RAND Corporation, 1986. As of February 20, 2009: http://www.rand.org/pubs/notes/N2505/

Wolf, Charles, Jr., Brian G. Chow, and Gregory S. Jones, *Enhancement by Enlargement: The Proliferation Security Initiative*, MG-806-OSD, Santa Monica, Calif.: RAND, 2008. As of February 19, 2009: http://www.rand.org/pubs/monographs/MG806/

Yamazaki, Mayuka, *Origin, Developments and Prospects for the Proliferation Security Initiative*, Institute for the Study of Diplomacy, Edmund A. Walsh School of Foreign Service, Georgetown University, Washington, D.C., 2006. Downloads as of February 17, 2009: http://isd.georgetown.edu/JFD_2006_PSA_Yamazaki.pdf